Techniques in Human Geography

A distinct menu of techniques is required for the study of topics in human geography. This book introduces the range of techniques now available to human geographers and explains how to use particular techniques within their appropriate context.

Techniques in Human Geography is the first concise guide to the purposeful use of techniques. Examining key techniques in detail – survey and qualitative, numerical, spatial and computer based – the book draws on important case studies, such as the decennial census, to illustrate applications. The importance of up-to-date IT-based techniques is particularly stressed, introducing widely recognised applications. A final section explores the Internet, which offers exciting new resources but also creates problems for researchers who are used to traditional academic fields.

Identifying important new directions of recent developments, particularly within computer mapping, GIS and on-line searching, Lindsay anticipates the ways in which techniques available to human geographers will change in the future, as well as the techniques which are available at the present.

James M. Lindsay is a Senior Lecturer in the School of Geography, University of North London.

Routledge Contemporary Human Geography Series

Series Editors:

David Bell and **Stephen Wynn Williams**, Staffordshire University

This new series of 12 texts offers stimulating introductions to the core subdisciplines of human geography. Building between 'traditional' approaches to subdisciplinary studies and contemporary treatments of these same issues, these concise introductions respond particularly to the new demands of modular courses. Uniformly designed, with a focus on student-friendly features, these books will form a coherent series which is up-to-date and reliable.

Forthcoming Titles:

Urban Geography

Rural Geography

Political Geography

Historical Geography

Cultural Geography

Theory and Philosophy

Development Geography

Tourism Geography

Transport, Communications & Technology Geography

Routledge Contemporary Human
Geography

Techniques in Human Geography

James M. Lindsay

London and New York

First published 1997
by Routledge
2 Park Square, Milton Park, Abingdon, Oxon, OX14 4RN

Simultaneously published in the USA and Canada
by Routledge
270 Madison Ave, New York NY 10016

Transferred to Digital Printing 2006

Typeset in Times by Keystroke, Jacaranda Lodge, Wolverhampton

British Library Cataloguing in Publication Data
A catalogue record for this book is available from the British Library

Library of Congress Cataloging in Publication Data

Lindsay, James M., 1947–
    Techniques in human geography / James M. Lindsay.
        p.   cm.—(Routledge contemporary human geography series)
    Includes bibliographical references and index.
    1. Human geography. I. Title.   II. Series.
    GF41.L55 1997
    304.2—dc21   97–7185 CIP

ISBN 0–415–15475–8 (hbk)
ISBN 0–415–15476–6 (pbk)

Publisher's Note
The publisher has gone to great lengths to ensure the quality of this reprint
but points out that some imperfections in the original may be apparent.

For Thomas, a geographer of the next millennium

Contents

Figures

⬤ Tables

Acknowledgements

I am glad to be able to take this opportunity to thank my colleagues and others who made this task a good deal easier than it might have been in many ways – rearranging timetables to clear time for writing, offering information, reading drafts, discussing key issues, lending books, and letting me make use of class time. I am particularly grateful to Jo Foord, John Gibbs, Jane Lewis, Helen Lindsay, Michael Mason, David Riley, Simon Tanner, and John Wareing. As a cartographer, John Gibbs's professional touch is also evident in those maps which did not come direct from the computer screen. Richard Burton provided technical support about the Internet and a host of other issues, and the staff of the University of North London's Information Support Service, particularly Oz Birch, did sterling work in solving the problems of temperament that machines display whenever they are called on to work under pressure. Alina Wiewiorcka of the University's Library service worked quiet miracles in locating errant books just when they were needed. I would also like to thank my students for acting as unwitting guinea pigs and providing a constant reminder of the difficulties of the learning and teaching processes. Last but not least I must thank Jean Tumbridge, long-suffering Department Secretary, for her endless supply of good humour and chocolate biscuits.

1 Introduction

Like all serious academic disciplines, geography has gone through turbulent times in the post-war years. Since the 'quantitative revolution' of the 1960s there has been a constant process of change in methodology. Geographers have developed or, more usually, imported new ideas. After being applied, evaluated, and criticised, these have found their niche or faded into oblivion.

Naturally enough the procession of changes in methodology has been accompanied by the development of different ranges of techniques. The unusually wide range of subjects in which geographers are involved means that geography as a whole has an unparalleled range of techniques at its disposal, ranging from the analysis of pollen assemblages, stream velocity, and atmospheric pollution on the one hand to participant observation of communities and the decoding of cultural landscapes as texts on the other. Renaissance geographers at ease with all of these techniques do not exist. We are all necessarily specialists.

Techniques have tended to be seen as the province of physical geographers. As a veteran of many funding meetings this author can testify that most of the money usually goes to buy equipment used in the physical area. Velocity meters and flumes, microscopes, and satellite stations for remote sensing might appear on the agenda. Big expensive bids for equipment for human geography will not. At one time it seemed as though the technical support needed for a project in human geography was no more than a sharp pencil. If we have moved on, it is by substituting a personal computer for the pencil.

There are grounds for this division. Research projects in physical areas often do require expensive equipment, and on the whole the costs of human projects tend to be measurable not so much in hardware as in time and labour, which do not come out of equipment budgets.[1] However, the equipment imbalance tends to create a false impression that human geographers somehow do not use techniques, reinforcing the impression that some of their physical colleagues already seem to have, that research in human geography is soft and lacking in academic rigour.

The importance of techniques for data collection and analysis in physical geography is probably the main reason why books on techniques in geography tend to be written for physical geographers, or at least emphasise physical examples. There are other reasons too, of course. Physical geography is more consistent in its use of the positivist approach discussed in Chapter 2, and this has created a body of well-defined procedures that any researcher in a particular field will be expected to adopt. It is both necessary and fairly easy to provide descriptions of these for their potential student users. Finally, it has to be said that newly developed techniques and the pieces of equipment that support them are exciting and rather glamorous, and their owners like to show them off as though they were new cars.

Although human geography has played its part in the development of quantitative techniques, the growth of a strong interest in qualitative methods has moved the emphasis into areas where defined procedures are relatively rare, and even in some cases unwelcome. It has to be said that there have indeed been textbooks about techniques for human geographers. In fact one of the earliest was published as long ago as 1971 (Toyne and Newby 1971). However, the number has been very small. On the principle that the devil should not have all the good tunes, this book offers its own contribution. It is an attempt to provide a guide for the 1990s. Anyone who doubts how far things have changed since Toyne and Newby published their book need only compare it with the on-screen tutorials provided by the excellent GeographyCAL software made available in the mid-1990s.[2]

Research is the basic focus of this book. At its most fundamental level research is simply a consistent way of asking questions and finding answers to them, and it forms an overarching framework which provides a home for all the techniques used by human geographers. Even elementary exercises should be seen in a research context. They are the first steps on a road that will lead to independent research in the form of a dissertation or project.

Although designed for students in the first stages of their academic careers, this text is also intended to look beyond the range of work that a first-year student is likely to carry out, providing a primary foundation for the more exacting projects that will be carried out in later years and directing its users to the sources that will be more appropriate at these levels.

We start by looking at the way that research is organised. Chapter 2 provides an overview of research design, particularly emphasising the importance of realism in choosing questions, working methods, and scales of operation appropriate for small-scale research tasks. Chapter 3 considers how we obtain, categorise, and assess the data we will need in our research work. Chapters 4 to 6 cover two very significant sets of techniques in research work in human geography, interview and questionnaire procedures and statistical testing respectively.

The emphasis in the remaining chapters is rather different. Having looked at techniques in relation to the research process we now turn to the hardware resources on which human geographers can draw in carrying out research tasks. Chapter 7 considers the maps and carto-graphic process which support all spatial description and analysis. Chapter 8 reviews the generic Information Technology fields most likely to be valuable in our context. All the remaining chapters focus on particular aspects of computer-based resources. In Chapter 9 we examine computer mapping and other specialised forms of geographical comput-ing, and this leads to a more detailed look at Geographical Information Systems in Chapter 10. Finally, in Chapter 11, we explore the Internet, which offers exciting new resources but also creates problems for researchers used to traditional academic fields.

It is not easy to achieve a good balance. If the sweep is too broad, there is a danger that the text will become simply a series of airy generalisations. If there is too close a focus on the mechanics of techniques, the text might become no more than a kind of cookery book of methods, the kind described aptly by one author as being 'as dull as a description of how to bang a nail on the head with a hammer' (Hakim 1992: 1). This book tries at all times to steer a middle course between these undesirable extremes. Its role is not to discuss the research process in abstract terms, and at the same time anyone looking for detailed instruction on any of the techniques described will find that the natural step forward is to move from this book to a specialised workbook or manual. Its concern is to map the range of possibilities in their research context, and explain *why* particular techniques are used in particular contexts rather than describe their working procedures in detail. It emphasises the importance of selecting techniques that will be fit and appropriate for their purpose, and being aware of their strengths and limitations.

Although geography has its own distinct emphasis – not least the concern with patterns in space – it shares a lot of ground with the social sciences, and there has been a good deal of interchange of ideas and methods in the last few decades. Like good Europeans, human geographers do not need passports to cross this particular frontier, and the text therefore moves freely through the relevant territories of social science as well as human geography.

Finally, we should note that we live in rapidly changing times. There have probably been few eras when this was not true of the human and the academic conditions, but ours can justifiably claim to be one in which the process of change has accelerated at rates far faster than ever before. Global transport, globalisation of the economy, and changes in the world's political balance have all changed the way we perceive and use space. It has been claimed that the development of the Internet might herald the growth of a genuinely spaceless society. Will geography have to reinvent itself?

One of the main themes of modern technology is convergence (Leslie 1996: 13). Information techniques that first developed in isolation are increasingly being integrated with each other. To take a humble but universal example, the domestic television set once sat alone in a corner of the living room. Now it finds itself playing host to video displays, cartridge-driven games, audio CDs and information on CD-ROM, photo-CD, teletext, satellite and cable input, interactive viewer response screens, and the Internet. The process of convergence itself is one of the important forces creating the changes we have just looked at. It also incidentally affects the media through which we study them. A book like this may turn out to belong to the last generation of paper-only texts.

Notes

[1] Things are changing. Although the grassroots perception may remain the same, the Research Councils and other research funding bodies now expect applications to be fully costed in all respects.

[2] Almost twenty GeographyCAL modules have been produced under the UK Teaching and Learning Technology Programme (TLTP). They provide useful and stimulating introductions to some themes in human geography as well as generic techniques discussed later in this text, and will be referred to where appropriate.

2 Research and project design – defining the context for the use of techniques

Students become involved in research very early in their courses, perhaps earlier than they realise. This chapter discusses:

- Methodology and research
- Research design
- Project design
- Data generation, analysis, and presentation

Techniques are tools, and all tools are designed with purposes in mind. If you buy a chisel then do nothing more with it than scratch your name on the table-top to demonstrate its sharpness, you have missed the point. In the same way geographers armed with all the skills that a technique can provide will waste their time unless they also knows the right (and wrong) way to use it.

The purpose that drives the application of techniques in human geography is the desire to undertake the research process. There is no need here for a long discussion of the meaning of the word 'research'. Authors who write at length about methodology and techniques are usually agreed that methodology provides the general framework used to approach the research topic, and techniques or methods are the mechanisms used to carry out the research task itself. On the other hand they often seem happy to leave the term 'research' itself undefined, as though we all shared an understanding of its meaning and significance. For our purposes here we can venture a definition of research as the process of systematically seeking answers to questions that are seen to be important by workers in a field. Posing questions and seeking answers is the bedrock of research, and this foundation is shared by all researchers, no matter what their position in the battle over methodology.

However there is a conflict, and it is impossible to stay out of it. Even if we do not take entrenched positions we have to recognise that choices have to be made. As we will see shortly, different areas of research have their own prevailing methodologies and techniques. Lines have been carefully drawn to distinguish between academic territories. However these battle-lines are not permanent. Fashions in research change, and the students of one generation may be encouraged to scoff at the excesses of zeal displayed by a previous generation in applying outmoded methods and techniques. The same fate awaits all fashions. The current of post-war thought has emphasised paradigms rather than unchallenged verities, interpretation, meaning, and symbolism rather than established facts. As the ground shifts underfoot it becomes increasingly dangerous to stand in the same place! The greatest wisdom in research is to be able to tell the difference between methodologies and techniques that are used because they are genuinely appropriate, and those which are used because that is what the research community currently expects. Awareness of this kind comes largely from experience.

The image of conflict is not meant to frighten the reader. A first-year student carrying out a straightforward exercise is not likely to be attacked physically by screaming research fundamentalists. The point is that all geography students are researchers from a very early stage and all research has a methodological context. The term 'researcher' might seem rather grandiose for people coming to terms with basic practical exercises for the first time, but all exercises are research projects in miniature, and have to be seen in this broader context. Preparing an essay or a seminar paper is a research exercise in its own right. A simple mechanical exercise in the use of a cartographic or statistical technique might offer you only a little window of choice and action in a framework which has been designed in advance. Nevertheless it carries a methodological loading, even though this may be heavily disguised if the project has not been imaginatively designed. The earlier in your student career you are able to grasp the context in which exercises and projects are offered, the better equipped you will be for individual research, above all for the dissertation that offers most geography students the best chance to display their skill in research techniques.

Research is a problem-solving activity, although a research 'problem' is not necessarily something that causes distress or difficulty. It is a question, an intellectual challenge concentrating on the need to acquire knowledge about an issue. Whatever research you are carrying out, it should not be difficult to identify the problem it is meant to solve, and if you cannot see what the problem is, the research may not be worth doing.

Research does not simply happen. All stages in a research project involve thought and decisions. The term 'research design' is commonly used to describe this process, but not always in the same way. You will find that some authors use it to cover the whole process of conducting a research project, including the formulation of research questions and the definition of a methodological context (e.g. Hakim 1992, Marshall and Rossman 1995). Others use it much more narrowly to cover the process of designing the data collection and later phases (e.g. Haring *et al.* 1992). In this text we will settle for a compromise. The term 'research design' will be used for the all-embracing strategic process, and 'project design' will be employed to cover the narrower tactical issues of putting the research design into practice.

Methodology and research

All research projects try to answer questions. Many research projects tackle questions by framing and testing hypotheses. A hypothesis is at its simplest a proposed answer to a research question. There will often be a great many possible answers, so the research process will involve creating, testing, and discarding different hypotheses until an acceptable one is found. However the stage of the research process at which a hypothesis appears, the way it is tested, or indeed the use of a hypothesis at all, all depend on the methodology the researcher is using. A major methodological issue we must look at before going further is the choice between inductive and deductive reasoning. Both forms of reasoning can be used to test and formulate hypotheses, but in quite distinct ways

Inductive reasoning, broadly speaking, works from the particular to the general. Observation of individual cases is used to propose hypotheses that will solve general questions. One form of inductive reasoning is the method immortalised by Sir Arthur Conan Doyle in the Sherlock Holmes stories, where the fictional detective builds up a composite picture from a series of individual observations to produce a striking solution to a problem. Taking a more mundane geographical example, we might want to find out which mode of transport students used to travel long distances. If we were working inductively we would survey users of all the main transport forms and use our findings to assess the preferences of all the student users we found. Where we use inductive reasoning it is inevitable that the hypothesis is not formulated until there is a body of data to work from.

Deductive logic works from the general to the particular, and uses general rules or laws to test individual cases. If we are using deductive logic we will set up a framework in which certain hypotheses are 'premises' from which conclusions can be drawn, e.g. in the form 'If

A and B, then C'. Returning to the student travel example, we might develop a structure like this:

- Coaches offer the cheapest long-distance form of travel (Premise A)
- Students always use the cheapest forms of travel (Premise B)
- Students travelling long distances go by coach (Conclusion C)

If testing of our hypotheses satisfies us that our premises are true, the conclusions too can be accepted as logically correct, and these in turn can be used as premises in further deduction. In other words, if we are able to accept our conclusion about student choice of transport, it can be treated as a premise in its own right. In this way deductive reasoning can set up complex chains or hierarchies of logic leading from initial premises to final conclusions. A research hypothesis normally performs a variation on this theme by defining the initial conditions then using evaluation of conclusions to test a hypothesis about the intervening state. It is normal to set out the hypothesis in advance of the process of data generation.

A rather more realistic example might clarify the issue. Suppose that we were doing research on the way that availability of transport affected the behaviour of shoppers. In a formal deductive structure with known laws we would use initial conditions about the shoppers (mobility, age, household form, for example) and a 'law' based on their willingness to travel, and then draw conclusions about their spatial shopping behaviour. In our research project, though, we can define initial conditions and measure the 'conclusions' in the form of shopping behaviour. It is the 'law' that we are trying to define and test. Our hypothesis might take a form such as: 'For a target group of shoppers, independent use of private transport for shopping trips significantly increases the spatial range of shopping.' We select the target group and measure the shopping behaviour, and this puts us in a position to accept or reject our proposed law about the effect of private transport.

In explanatory terms inductive reasoning has one major problem. However many instances we find that support our hypothesis, we cannot be sure that we have missed others that might disprove it. A hypothesis based on a huge mass of inductive evidence might be invalidated if we find just one case that behaves quite differently. Deductive structures can also be invalidated by instances that break their rules, but methodology has found a way of converting this to a strength. As long ago as the early 1960s Karl Popper developed the doctrine of falsificationism, which is based on the belief that no hypothesis can be proved, and that science can advance only by proving successive hypotheses false. From this emerged the grandly named 'hypothetico-deductive method', in which hypotheses are tested to see whether or not they can be disproved. In the case of our student travel example, if we find that all our surveyed students travel by coach, we can uphold the hypothesis at least for the time being. If we find one individual travelling by rail or boat, we must discard it and find an alternative.[1]

This is not the place for a prolonged discussion of methodology in human geography. That can be found elsewhere. What is important here is to see how the argument about methodology has affected research in the range of fields that comprise human geography. Broadly speaking research in human geography is shared rather uneasily by two different methodological traditions, one inherited from the scientific tradition in geography and the other acquired from its relationship to the social sciences. One of the few consistent features has been polarisation, with individual geographers tending to align themselves with one or other approach and committing themselves to a methodology that confirmed the correctness of this choice.

Table 2:1 provides an outline of the significant ways in which the two main approaches differ. The table is based with modifications on Sayer's summary (Sayer 1992: 243), which is itself developed from the work of Harré (Harré 1979: 132–134). It uses Harré's terms 'intensive' and 'extensive' to distinguish the two forms. The geographer's habit of referring to 'quantitative' and 'qualitative' approaches focuses on the difference between techniques which evaluate variation in numerical value and those that do not.

Extensive research is research in the 'scientific' tradition. It is 'positivist' in the sense that it tries to establish general laws using quantitative methods, usually in the hypothetico-deductive framework described earlier. Table 2:1 summarises its characteristics. Because it is concerned with laws and regularities, research questions are typically about populations rather than individuals, although we almost always work from representative samples rather than total populations (see Chapters 4 and 5). We work taxonomically by classifying people into groups in some appropriate way. In the shopping example we try to classify people into groups with different shopping behaviour. The relationships we are testing are associations. This is a very important point. We are examining people's behaviour as members of groups and not as individuals, and individual motivation is not being explored. Typically we work with relatively large numbers of cases, and statistical tests will often be performed to test the hypothesis. However, even if we can demonstrate very convincingly that a relationship exists, we do not know that difference in access to transport *causes* difference in shopping behaviour. All we know is that for our group the two characteristics are associated at some level of statistical significance (see Chapter 5).

Table 2:1 *Extensive and intensive research methods*

	Extensive	*Intensive*
Research question	What regularities are there in a population's behaviour? How characteristic of a population are the studied processes?	How does a process work in a particular case? What is the cause of a particular change? How does the process actually work?
Relationships	Associations	Connections
Type of group studied	Groups linked by taxonomy	Groups linked by causality
Typical methods	Large-scale surveys of populations or samples Formal questionnaires Standardised interviews Statistical analysis	Study of individual agents in causal context Interactive interviews Ethnography Qualitative analysis
Verification	Replication	Corroboration

The classic method of hypothesis formulation starts with a research hypothesis derived from a theory under test. We will almost always be able to test this only on a sample of data (see Chapters 4, 5 and 6), and our primary task is to establish whether the sample's behaviour represents the behaviour we could expect from the population as a whole. The actual testing involves the formulation of a null hypothesis (symbolised as H_0), which proposes that the relationship visible in the sample set of data does not represent a relationship in the whole population. This null hypothesis is set up in the hope that it can be rejected, and we will also set up an alternative hypothesis (H_1), which is that the sample data is in fact representative. This may seem circuitous, but it provides a suitably cautious basis for progress. Our primary aim is to reject H_0. If we are able to do so, we can accept H_1 although we cannot claim to have proved it. We are simply upholding it for the time being without denying the possibility that other better alternatives might be found. If H_0 has to be upheld, of course, we have to discard H_1 and start again.

An important issue emerges here. The hypothesis-testing process does not give us access to absolute truth. When we accept or reject a hypothesis we do so as the result of a process of evaluation, and in

Table 2:2 *Types of error*

	Null hypothesis true	Alternative hypothesis true
Reject null hypothesis	Type I error	Correct decision
Accept null hypothesis	Correct decision	Type II error

consequence we can never be absolutely certain that we have made the correct decision. Table 2:2 shows the results of rejecting or accepting H_o in different circumstances. A Type I error takes place if a verifiable null hypothesis is rejected, and the alternative is wrongly accepted. A Type II error means that an acceptable alternative is rejected and the null hypothesis is upheld. On the whole a Type I error is normally taken to be the more serious, since it upholds a wrong alternative. A Type II error discards an acceptable explanation, but at least does not propose a false alternative.

Unfortunately the assessment of seriousness is a value judgement, as Harvey points out, and so is the degree of caution we exercise statistically about the possibility of a mistake (Harvey 1969: 252). We could only be sure which of the four outcomes shown in Table 2:2 was correct if we possessed an unattainable state of perfect knowledge. In real life the selection of the 'correct' outcome is a matter of judgement. Even if the full power of statistics is available, the researcher has to be aware that the choice of significance level (see Chapter 5) is critical. There is a continuum from a high risk of Type I errors and low risk of Type II errors at the lax end of the significance scale to low risk of Type I and high risk of Type II at the strict end. Sadly there is no certainty about the correct point to choose between these extremes. Type I errors are more likely with large samples and Type II errors if the sample is small. For this reason most texts recommend as a rule of thumb the use of different significance levels for different sample sizes, but the judgement about this is conventional and arbitrary (de Vaus 1996: 191).

Since hypothesis formulation takes place early in the research process, a great deal of planning has to be carried out before any data generation takes place, and the validity of the findings depends on careful adherence to the planned structure. Improvisation and unscheduled responses are definitely not encouraged. If a hypothesis is upheld, it should be capable of verification by 'replication'. In the controlled situations of the physical sciences, research findings can be tested by trying to produce comparable results at will using the same methods, even in different laboratories and at different times. Most human geography does not happen in laboratories but in complex human environments, and it has always been recognised that true replication is only rarely achievable. The best the researcher can usually do is to control conditions to minimise the number of variables involved, and use statistical techniques to assess the confidence we can attach to the outcomes.

Intensive research is very different. We can describe it quite briefly, because its emphasis in different fields is consistent and intuitively understandable. It is essentially qualitative, seeking to understand individual perceptions rather than collective behaviour. As the name suggests, the emphasis in intensive research is on individual or small-scale work, actively pursuing causality and the range of response. Every section of the Intensive column in Table 2:1 reflects the same concern with individual behaviour and causal relationships. This is a fundamental difference. Even when individual behaviour reinforces general rules it may be prompted by distinct motivations (for example, people do not start shopping in superstores in order to concentrate control in the hands of a few retail chains, even if that is the result of their actions). Reasoning is usually inductive rather than deductive, and although there will always be research questions or problems, not all researchers use formal hypotheses. The research methods favoured in intensive research emphasise observation, interaction, and response to individual cases. An interactive interview will follow whatever path seems best in context, whereas the standardised interviews or questionnaires of the extensive mode cannot deviate from their prescribed forms. It should be obvious at this point that replication can have no place in an intensive framework. The only way of verifying or supporting findings in this field is by corroboration, making use of other information about the individual research subjects to ensure that the research findings are really applicable. However there are distinct forms of qualitative research and we can appreciate the range by looking at two of the main schools.

Interpretivism treats human behaviour as a text to be analysed, and this analysis or interpretation is made both by the researcher and (consciously or unconsciously) by the social actors involved in the study. A common understanding among interpretivists is that the researcher cannot stand objectively outside the context of the study. In these terms an interview is a 'co-elaborated' act rather than a formal exercise in data transmission and collection, and a transcript from such an interview is a script which will lose much of its meaning if any attempt is made to summarise it. This very heavy emphasis on the particular and the unique means that researchers in some schools of interpretivism deny the value or indeed the possibility of formulating general laws.

Ethnography on the other hand is in the narrow sense a descriptive method involving careful observation and recording of social behaviour, often over long periods. It is essentially the approach of social anthropology, in which the researcher always remains an external observer, even if after some time the social actors barely notice the researcher's presence. Ethnographers generally do not share the reluctance of interpretivists to risk losing the individual meaning of texts like interview transcripts by condensing or coding them, and they are more likely to be ready to construct and test theories (Miles and Huberman 1994: 8). However we should not make too firm a distinction. Followers of Clifford Geertz, who has had a great deal of influence among cultural geographers, see anthropology as an essentially interpretative science looking for meaning rather than an experimental one looking for laws. From this perspective ethnography too is seen as an interpretative process (Jackson 1989: 172–173).

The differences (and the overlaps) between these approaches show that it would be a mistake to treat intensive research as a unified field, or indeed one dominated by deliberately unstructured approaches. However, where it does use structures they have a quite distinct role. In extensive research, structures in the forms of hypotheses are erected at the start of the research project. Theory-oriented intensive researchers are more likely to build structures of explanations in the later stages of analysis on the basis of their findings, and a whole range of techniques has been developed to allow conclusions to be drawn and verified with academic rigour (Miles and Huberman 1994: 11).

The extensive and intensive approaches have distinct strengths and weaknesses. Extensive researchers have traditionally been proud of their objectivity, elimination of observer-induced bias, and academic rigour. On the other hand the emphasis on general principles and the behaviour of populations means paradoxically that the power of explanation is narrowly focused. In other words, because there is no real attempt to explain individual behaviour or patterns of causality, large amounts of potential data are discarded or ignored, and its critics have regarded it as an impoverished mode of explanation. Intensive research, in turn, has problems which stem from its emphasis on the particular and lack of any real means of generalisation. Even theory-oriented researchers find that there is a lack of well-formulated methods of analysis and clearly defined conventions, and some intensive researchers have reacted against the perceived excesses of the extensive approach by coming to regard a lack of formal academic rigour as a virtue in itself. Because of this, even some supporters of the approach have felt it necessary to criticise a tendency to play host to anecdotal, subjective, and unrigorous research (Silverman 1993: 43–44).

Both approaches therefore have a place, and that place is determined by appropriateness. In a rational world researchers would be able to make free use of either methodology, and it is unfortunate that in the whole area of human geography and social science their use has become polarised, with researchers in particular fields tending to become proprietorial about their own research methods and dismissive of others. There is no shortage of voices pointing out that these methods should be treated as complementary and certainly not mutually exclusive (e.g. Silverman 1993: 21–23, Sayer 1992: 244–251), and in the social sciences some of the barriers may already have begun to come down (Miles and Huberman 1994: 4–5). It is a pity that researchers do not always listen and respond. The novice researcher should certainly feel free to explore the range of methods without feeling obliged to apply for life membership of either of the tribes.

Research design

So far we have looked at general issues. We have dwelt on them at some length because a proper appreciation of methodology and techniques is the sound foundation on which good research is built.

Now we come to the important question. How do you design your own research project? A lot depends on its nature and scale. Smaller research tasks like class exercises and seminar presentations will usually give you very limited freedom, but at the other extreme a dissertation gives you responsibility for taking a research project all the way from conception to completion. In the text that follows we will look forward to the stage when you have to plan a dissertation, but that does not mean that the chapter should be left unread until that stage. Most of the areas we look at will be just as relevant in a scaled-down form to smaller exercises. Quite a lot has been written already about dissertation and project design, tackling the topic in more depth than can be provided here. A well-established guide to projects in social science is Judith Bell's *Doing your Research Project* (Bell 1993). A more specific guide is *How to Do your Dissertation in Geography and Related Disciplines* by Tony Parsons and Peter Knight (Parsons and Knight 1995), although it emphasises physical topics in particular. A number of departments produce their own dissertation guide covering working principles as well as local regulations, and you should obviously use one if it is available. Table 2:3 shows the stages through which a research project is likely to evolve.

All research projects emerge from an interest in a particular topic. A broad aim is defined from the start. Sometimes the research question

Table 2:3 *Stages of a research project*

Choice of research topic	
Literature review and contextual study	
Choice of research design	Choice of methodology and research approach
	Definition of research questions [may feed back into previous stage]
Project design	Logistics and timetabling
	Assessment of resource needs
	Liaison and permissions
	Pilot study
	Revision of project design if necessary
Project implementation	Data gathering
	Data analysis
	Interpretation
Presentation	Text
	Supporting materials

itself will be clear from the earliest stages, but in other cases it will only emerge after a review of the literature in the field. A literature review is a vital early part of any serious research project, and is basically an exercise in context definition. Carrying out the review lets the researcher identify the kind of research being done in the field and get an overview of the methodological and technical frameworks that have an established place in the field.[2] It will also help identify areas where academic debate is going on and research questions that are both live and interesting. In the dissertation context the literature review will have a formal place in the text itself, allowing you to show how your work fits into the broader context of research in the field.

Finding the literature is not always an easy task. There are so many outlets for published research work that a shelf-by-shelf search of your library would be both tiring and inadequate. Recommendation can help. A specialist might be able to advise you about important recent publications or conference proceedings. Scanning the indexes and bibliographies of recent publications by workers in the field can also be useful, although publishing is a slow process and even newly published books and articles are likely to be a year or more out of date.

One of the most effective ways of scanning the literature is to use an abstracting system. The longest-established general abstract series in the geographical sciences is Geo Abstracts, which provides a large-scale summary of book and journal publications broken down by major subject areas. Each entry provides details of the publication and a short abstract summarising its contents. In the mid-1990s this system took on a computerised form on CD-ROM as GEOBASE, which most university libraries now support. GEOBASE, which is revised quarterly, derives its material from about 2,000 journals and other sources in geography and related fields, and it provides a much faster and more flexible search system than its predecessor. However it shares the drawbacks of all abstracting services. Because someone has to read every item and write the abstract an element of delay is inevitable, and there is no guarantee that your library will have direct access to every item you might want to read. It also has to be said that a variation of Murphy's Law ensures that the references with most potential somehow always turn out to be unpublished conference papers written in minority languages.

Having digested the literature and surveyed conditions in the area where you will carry out the fieldwork, you should be able to define the aims of your research work more precisely and refine the research questions. This is a feedback process, in which every revision of the aims can open up new areas of literature, and this might mean further modification of the aims and hence the reading of a different literature. When the dust has settled it is time for the research questions to be framed, even if we are aware there might have to be minor revisions at the next stage when the logistics of the exercise are reviewed. The research design is complete, and now we can turn to project design.

Project design

It is not enough for a research project to be interesting and academically acceptable. It has to work. Student projects suffer more than postgraduate research from the constraints of time, access to data, and resources. It is important to find a realistic scale of operation. If your project is too ambitious to work within these constraints it will inevitably fail to match its potential and might end as a heroic failure. On the other hand a 'safe' project that operates well within the constraints may finish as a disappointing and uninformative piece of work. There is no simple formula to determine what the limits on your scope might be. The best policy is to be as thorough and systematic as possible in exploring in advance the problems you might meet.

Time is a major constraint. Student projects have to operate within periods of weeks or months. There can be no scope for long-term studies.

You should be aware that if the topic you are working on varies seasonally, you will at best be able to collect data for one sequence of seasons. You also have to be careful to relate the events you are studying to your own working timetable. The dissertation cycle typically starts in midsummer and data should have been collected by the end of the autumn. If you choose to study the impact of winter sports on a tourist host community the data collection season will be dangerously late and might not give you time to complete a full analysis before the deadline. Building a timetable that gives a realistic weight to the different stages and relates them to other demands on your time is essential.

You should also be confident that you can get access to data. Questionnaire fieldwork in city streets might not need permission but fieldwork often has to be done on private property. If you need permission in advance, get it! Many property owners and organisations are ready to give authorisation in advance, but they may take a hostile line to students found working without permission, and this could leave your data collection disastrously incomplete. Some organisations (including the managements of major private shopping centres) have a general policy of refusing permission to student researchers. If your dissertation relies on data from official bodies you should negotiate access well in advance to ensure that bureaucracy can process your request at its own speed. The price of their co-operation may be a courtesy copy of your finished work, a fairly small matter in these days of word-processing. A letter of support from a supervisor or Head of Department is a traditional way of establishing your credentials with data owners of all types.

You should also assess your own resources. Elementary as it might seem, some student researchers fail to budget for the costs of travel, accommodation, and fieldwork. You should also make arrangements as early as possible for maps and field equipment, any field assistance you might need, and support in production of the final report. Skill is also an important resource. Only an unwise student would embark on a project that required specialised techniques in which he did not have training in advance. This is a more common problem in physical geography, but a surprising number of human projects come unstuck at the analysis stage because of a failure to realise in advance that statistical techniques require careful handling (see Chapters 5 and 6).

One of the best ways of minimising problems is to carry out a small pilot study. This will help point out unforeseen problems in the data generation and handling process, and the cost in terms of time and effort should be outweighed by the improvement to the full study. Pilots are specially important for questionnaire-based research (see Chapter 4), since you will not be in a position to redesign and print a new questionnaire if the first day's fieldwork reveals flaws. As a general rule teething troubles mean that the first day of field data collection is the most difficult, and even if no other kind of pilot study has been possible this day can be treated as a kind of mini-pilot.

Data generation, analysis and presentation

The heart of any piece of research work is data generation, analysis, and the interpretation of results. Success in these fields will create a strong piece of research needing only good presentation to make a major impact. However these are the topics about which least will be said here. There are two good reasons behind this apparent paradox. One is that these areas are the most specific. Every project will use its own type of data in its own way defined by its aims and hypotheses, and the relationship between aims and data will determine the choice of analytic method. The contrast between extensive and intensive types of research discussed earlier should have made it clear how wide-ranging the forms of data handling and analysis might be. The other reason is that the chapters that follow will pay attention to major issues relating to data generation and analysis, and they do not need to be rehearsed here. Questions of data type, quantity, and evaluation are taken up in Chapter 3. Issues relating to interview and questionnaire analysis are discussed in Chapter 4 and statistical methods in Chapters 5 and 6.

One important point must be made at this stage. Even a meticulously designed project complete with pilot studies may throw up unforeseen problems during these stages. If changes have to be made in working method, these have to be fully recorded. Even if the techniques used allow it, there will probably not be time to revisit the study site and make good any inconsistencies. It is vital to be consistent and to make sure that alterations to working methods are fully documented. This is important not only as part of the project in itself but as a record of the learning process. Unless they are documented, late alterations of this kind can easily be forgotten before another project is launched.

Every research project of whatever size will have to conform to some set of rules of presentation. These are discussed in general terms in the dissertation and project texts referred to earlier, and very specific requirements might be laid down for particular projects. This is the stage at which the project goes through the series of drafts which lead to the production of the final text or report. Writing will take longer than you expect, and it is a very rare first draft that is fit to go forward as a final statement. Graphics, tables, and other supporting material will have to be produced and checked. Even quite short research reports will also need supporting material like title pages, contents lists, and reference lists. A failure to attend to these details may reduce the weight of the substantive findings of your research.

Summary

One of the most important things to learn about the research process is the need to be realistic about the scale of your work. Be careful to ask research questions that can be tackled practically with the resources you have available. Constraints of time and resources mean that very few student projects or dissertations can be treated unreservedly as major pieces of research. It is a very rare dissertation indeed that can be recommended for publication as it stands. You should be aware of the limits to your scope and aim to do your best within them. Over-ambitious dissertations may gain respect of a kind, but they are unlikely to do themselves justice. The best student research is done by people who recognise the limitations within which their work operates, but can use it to demonstrate what they might be able to do later on and on a bigger stage.

Discussion questions and tasks

1 Just when you thought it was safe to go back in the water. . . . There are other important and relevant terms we have not used here. Find out for example about structuralism, semiotics, the relationship between ideographic and nomothetic approaches, and see how they relate to the terms we already know.

2 Take a volume of a geographical journal – a broad-spectrum journal like *Transactions of the Institute of British Geographers* will serve well – and do an inventory of the methodologies and research designs that emerge from its articles.

3 You have been asked to carry out research on the mapping of deprivation in British inner cities. Sketch out a project design, anticipating which issues will need special attention.

4 Do the same thing in the context of the effects of desertification on the movements of nomads in sub-Saharan Africa.

Further reading

See also

Data, Chapter 3
Questionnaires and interviews, Chapter 4
Statistics, Chapters 5 and 6

General further reading

A browse through the following books will give some guide to the range of methodologies in Geography and the Social Sciences.

Eyles, J. and Smith, D.M. (eds) (1988) *Qualitative Methods in Human Geography*, Cambridge: Polity Press.

O'Brien, L. (1992) *Introduction to Quantitative Geography*, London: Routledge.

de Vaus, D.A. (1996) *Surveys in Social Research*, London: UCL Press, 4th edition.

Denzin, N. and Lincoln, Y.S. (eds) (1994) *Handbook of Qualitative Research*, Thousand Oaks: Sage.

The two texts below provide a good guide to sound practice in designing and carrying out research projects.

Bell, J. (1993) *Doing your Research Project*, Buckingham: Open University Press.

Parsons, A. and Knight, P. (1995) *How to Do your Dissertation in Geography and Related Disciplines*, London: Chapman and Hall.

Notes

[1] For illustrations's sake, this example has been made unrealistically rigorous. Research workers in human geography normally work with probabilities rather than exclusive conditions. See Chapter 5.

[2] Without prejudice to your own choice based on appropriateness! See the discussion earlier.

③ Handling data

All research exercises use data. This chapter covers:

* An approach to the classification of data
* Qualitative and quantitative data
* Types of data
* Evaluating data

Data[1] are the materials from which academic work is built. As such they are ubiquitous. From passenger counts on transport systems to the constructs used in the most abstract discussion, data always have a place. However it is important to recognise that like other building materials, they have two universal characteristics. The first is that they are (or at least should be) selected on the basis that they are fit for their intended purpose. Data generation is a purposive activity, and must be governed by the task in hand. This means that data can lead multiple lives, sought after for some purposes and spurned for others. The second point is that they are not unmodified products of nature. Data are not only selected for the purpose, they are shaped for it.

The language used to describe the process by which we get data might suggest otherwise. The term 'raw data' is still commonly used. Although it has its value in distinguishing the primary data used in a project from secondary refined forms, it conveys an unfortunate impression of data as winnable commodities like potatoes in a field or iron ore fresh from the mine. The familiar terms 'data gathering' and 'data collection' also create an image of the scientist harvesting natural products, basket in hand. It is a long time since most scientists were comfortable with the image of themselves as empirical searchers for nuggets of revealed truth, and there have naturally been attempts to find alternative terms. Among social scientists, often much more directly conscious of the role of language than geographers, several alternatives have been suggested. The terms 'data making', 'data production', and 'data construction' have all had their advocates (Bateson 1984: 5, Sayer 1992: 272n). They have also had their critics, since they all suggest an element of conscious manufacture which runs contrary to the accepted ideal of the scientist as a transparent medium through which the truth is elucidated, and at the worst imply downright falsification.[2] 'Data generation' is perhaps the best choice from a poor field, on the one hand avoiding the danger of suggesting that data are raw natural products, and on the other not over-emphasising the element of manufacture.

This is not a trivial point. If data are not naturally given objects, then we have to recognise that they are chosen and shaped in a way which may have unconscious as well as conscious elements. Research in the perception field has emphasised that the processes of observation which underpin data generation are governed by our systems of concepts, and the data we obtain are already pre-conceptualised even before they are subjected to the dangerous process of analysis and interpretation (Sayer 1992: 52). This casts yet more doubt on the already beleaguered concept of scientific objectivity, and in an academic world where post-structuralism has worked hard to demolish the foundations of any system of explanation (including its own), the student might reasonably ask whether any traditional model is still appropriate. The answer in this case must be a qualified 'yes'. In the absence of a convincing new paradigm, we can do little better than carry on with traditional methods, but at least with a better awareness of their weaknesses.

An approach to the classification of data

Data can be classified in several ways. Most of this chapter will be devoted to looking at data in terms of their intrinsic characteristics and the ways in which they can be manipulated. However, we will start by looking at data in terms of provenance. Data can be classified as primary, secondary, or tertiary, according to the source from which they come.

Primary data are those collected directly by the researcher for a particular purpose, and it is an unusual research exercise that does not generate its own. Although the process is often time-consuming and laborious, it has the advantage of giving the researcher as much control as possible over the data used. In principle at least primary data have not been filtered, edited, interpreted, or subjected to any form of unstated bias by external agents. Although primary status does not confer a magic cloak of objectivity on data, it does at least mean that they approach as near as possible to the ideal state. Typically these data will have been collected in the field or laboratory. From the human geographer's point of view, the 'field' is most often a street or public meeting place. Some topics will involve measurement of objects of phenomena, but a great many will involve people individually or collectively. Questionnaires and interviews are therefore important sources. As for the 'laboratory', this for a human geographer is most likely to be a computer room or data archive.

No research task exists in a vacuum, and there will always be a role for secondary data. These have inevitably been processed in ways that primary data have not. Traditionally, one role of secondary data is to provide a context in which the primary materials are handled and analysed, and a perspective in which the analysis is viewed. Because it has been filtered and subjected to judgement, a research paper containing the primary data of one study will provide one of the secondary sources for another. However the range of secondary sources is much wider, including newspapers and other current affairs sources and books published outside the academic area. Broadly speaking, primary sources analyse fresh data and secondary sources provide us with the results of a single filtering of primary data. The concept of a tertiary level is useful in identifying a context for textbooks, which typically integrate and discuss secondary studies, and as such are a further level removed from primary data. However the distinction between secondary and tertiary data is not always easy to make, and we should not concern ourselves too much with it. The distinction between primary and secondary data is the one that matters.

At the cost of perhaps complicating the picture a little, it is worth noting that there are special cases. Large-scale collections of statistical data, typified by national censuses, do not qualify strictly as primary data. Their data have inevitably been collected on the basis of processes outside the user's control, and are often quite highly structured. As we will see later, data values may even be falsified under controlled conditions to protect confidentiality. However, it is normally possible for the researcher to get a good understanding of the rules under which the data have been processed, and the processing of most national censuses is free of overt ideological or political intent. For these reasons, data sources like these are treated as primary.

It is also possible for a researcher to use unashamedly secondary or tertiary material as primary data if the context allows. Two examples should make this clear. Nineteenth-century travellers' journals are on the face of it very weak data sources. They often consist of tangled mixtures of primary observation and unverified secondary material, written in a literary style rife with both overt and implicit value judgements. Their authors were also free to embroider their tales in the knowledge that few of their readers would ever be in a position to check them. From the point of view of the historical geographer, they are secondary sources to be used only with great care. However anyone carrying out research into the changing perception of landscape and culture is quite justified in using these texts as primary sources. In this context the value expressions which make them unsuitable for other purposes become primary data in their own right. Similarly, a student of the development of approaches to regional geography would be fully entitled to use textbooks of regional geography, tertiary sources in normal circumstances, as sources of primary data. If we are looking at the ways in which authors view the regional issue and select data to fit their perceptions, even the stodgiest of elderly regional textbooks become primary sources in themselves.

Qualitative and quantitative data

Data can of course be categorised, and one of the most fundamental ways of doing this is to make the distinction between quantitative and qualitative forms. Quite simply, quantitative are distinguished from qualitative data as those which can legitimately be analysed using numerical techniques. Sadly this simplicity is entirely superficial. The relationship between the two forms has become part of the much larger dialogue between ideologies and methods of explanation, quantitative and qualitative data being the characteristic forms of extensive and intensive research respectively. We have already looked briefly at this ideological argument (Chapter 2), but it is worth thinking of its significance in the context of data.

One of the concerns of the movement towards quantification that took place from the mid-1960s onward was to improve the academic standing of the field by putting it on a methodologically firmer footing. To the revolutionaries of the time geography was seen (not without cause) to be dominated by weak methodology with an emphasis on the accumulation of fact, anecdotal data handling, and explanation that tended to be intuitive and inductive. It is hardly surprising that their new approach to methodology was reflected in this assessment of data. Committed quantitative geographers came to regard qualitative data as little more than an unsatisfactory rump that could not be subjected to a recognised form of treatment, or at best a 'soft' resource used in the exploratory stages of a study. An emphasis on the use of deductive reasoning and numerical techniques to test hypotheses was seen as necessary to bring geography into line with conventional scientific method. The experience was not totally happy.

The example of historical geography demonstrates some of the difficulties. By their nature, the data available to historical geographers have suffered over time and tend to be fragmentary and scattered. This encourages squirrel-like collection of whatever can be found, and in the circumstances systematic hypothesis testing is very difficult. Pre-quantitative historical geography was notoriously the home of an unstructured form of inductive reasoning, which was often ingenious in the way that arguments were developed but uncritical about the data used to support them. Progressive historical geographers quite justifiably tried to improve the standing of their field by adopting a rigorously quantitative and deductive approach, but they did not always get it right. Some, like David Harvey, later admitted that in their pioneer zeal they had used techniques wrongly.[3] Others did not! A more general problem was that the limited availability of data drastically narrowed their scope for study, forcing them to work on the relatively rare topics which could offer consistent data, and this often meant that they were asking trivial questions and producing conclusions that were strikingly banal.

Similar polarisation happened in the social sciences. Here too some quantifiers came to regard themselves as the sole custodians of academic rigour. This self-consciously tough approach was criticised some years ago in a review under the ironic title *Real Men Don't Collect Soft Data*.[4] The qualitative schools have tended to denounce quantification as the triumph of technique over purpose. The quantifiers, epitomised by survey analysts, have regularly been accused of taking a soulless approach, using a form of mechanical number-crunching which was inappropriate in the interpretation of the subtle dynamics of social interaction. As was argued in Chapter 2, polarisation of this type is neither necessary nor productive. Apart from anything else, it means the loss of flexibility in explanation. Silverman has denounced it vigorously in the context of the social sciences, arguing that the choice of method should be determined by appropriateness to the topic. Quite simply, 'it all depends on what you are trying to do' (Silverman 1993: 22). Geographers would do well to be aware of the possibility of using more than one approach. As we saw in Chapter 2 the research question itself should drive the study in the majority of cases. Only when the question is defined should issues of methodology and data be tackled.

Types of data

All data are obtained by some sort of measurement, although at its simplest the measurement might amount to no more than allocation to a class. Every form of data can be assigned to one or more types, and the form that this measurement takes will depend on the type of data being collected. There are a great many possible data types. Anyone designing fields to store data in a computer database manager might have to choose from a menu of fifteen or more different types. In reality this worryingly large range of choice can be trimmed drastically. Some of the types defined in database contexts have no relevance outside the computer. These include numerical types defined in terms of their special computer storage requirements, and others like currency where the only distinctive feature will be a currency symbol displayed on the output. In essence only three of the computer-based types have any claim to be universal. These are alphanumeric, numeric, and structured. The emphasis here on computer forms is deliberate. Most data handling computer applications are type-bound to some degree, in the sense that the user is required to make a decision about the way in which data are classified. The discipline forced on the user by this can help clarify some of the more ambiguous aspects of the storage of data on paper.

It is hardly necessary to define the numeric category. However it is worth emphasising that numeric data will normally be either discrete ('integer' in computer parlance) or continuous ('real'). A retail geographer studying shopper behaviour in a shopping mall, for example, may well be recording and analysing visitor numbers and spending behaviour. The primary data for shopper numbers will inevitably be based on discrete counting – shoppers do not come in half-quantities – but information about spending is likely to be measured as continuous data in fractional currency sums. Alphanumeric data consist of text, ranging from single letters or characters through to large blocks of text, or 'strings' in computer terminology. Numbers can belong to either category and thus lead a double life. However there is a great difference between the functions of a number in its two different incarnations. This is illustrated well by our third type, structured data. Common examples are formatted dates, telephone numbers, and any form of standardised personal identity number. Typically these are organised into groups of digits in consistent forms. Although frequently entirely numerical in appearance, there is no realistic way that arithmetic can be performed on them. For this reason, structured types can be regarded as a special form of alphanumeric data.

If we step back outside the computer context, we would normally include structured types in the alphanumeric category, and we are therefore left with a simple two-fold classification into numeric and alphanumeric. Broadly speaking, this division matches the distinction into quantitative and qualitative forms which we have just examined. However there is some overlap in the range of techniques that can be applied to them.

Alphanumeric data can be used quantitatively, as we can see if we look at the hierarchy of scales of measurement.

Traditionally numerical data handling makes use of four scales of measurement, which make up a hierarchy of increasing power. These are shown in Table 3:1, and examples are illustrated in Table 3:2.

Nominal data consist of items which are categorised only as types within a series. They may have attributes or attached values that can be processed mathematically, but they cannot be processed themselves. The first column in Table 3:2 lists the seven main categories of local authority spending as they were defined in London in the late nineteenth century. These are names and no more, although they are consistent in being members of one group. In this particular instance we have only the single occurrences of the categories for the case 'London', but these categories and possibly others would appear in the budgets of other British cities of the time. The order in which they are listed here has no mathematical significance, and indeed the names themselves can be changed without affecting their role in the series. 'Parks and Open Spaces', for example, could as easily be called 'Open Spaces and Parks'.[5] The only numerical operation we can perform on nominal data in themselves is calculation of the mode, the value which occurs most commonly in a group. A survey of several cities might show that the mode, i.e. the category most often encountered, was 'Main drainage'. This gives us very little extra information. All we can learn about the modal category is that it occurs frequently, but we cannot tell whether the values attached to it are consistently large or small, important or insignificant.

Table 3:1   *Scales of measurement*

Scale of measurement	Arithmetic potential	Data type
Nominal	Mode	Both
Ordinal	Median	Both
Interval	Mean	Numeric
Ratio	Mean	Numeric

Table 3:2   *The London County Council's capital expenditure, 1889–1896*[6]

Category	Rank	Total expenditure (£)	Expenditure in relation to mean
Street improvement	5	577,770	–64,228
Main drainage	2	884,084	242,086
Parks and open spaces	4	584,538	–57,460
Bridges, tunnels, ferries	1	1,186,405	544,407
Clearance schemes	6	360,002	–281,996
Artisans' dwellings	7	180,401	–461,597
Lunatic asylums	3	720,786	78,788

Ordinal data are those which can be ranked or assigned to a place in a series. One way of doing this is to put values into a framework of ordered categories. We could use the total expenditure figures in Table 3:2 to devise a simple set of categories, perhaps 'High', 'Medium', and 'Low', within which the values could be classified. However we can get more from our data by ranking them individually. In Table 3:2 the seven categories are ranked in accordance with known spending. Here as it happens the rank is derived from comparison of absolute values, but rank will often be established by assessing relative speed, height, proximity, or some other quality without the benefit of a precise measurement. Arithmetical processing becomes a little more advanced here. It is possible to define the median, the value which occupies the mid-point of the scale. In this case the fourth-ranked category, 'Parks and Open Spaces', is the median. However the level of information is not much more. If the values are distributed in a grossly skewed way with the majority at one end of the numerical range, the median will tell us little of value about the distribution. Rank values cannot in any case be added or subtracted in any arithmetically meaningful way.[7]

Interval and ratio data have a lot in common, and in fact can usually be treated in the same way. Both are based on scales which are defined in terms of equal units. Depending on context, both discrete (as in Table 3:2) and continuous forms are legitimate. The distinction between interval and ratio data is that ratio scales have an absolute or natural zero point. A measurement of weight in kilograms or distance in kilometres is thus on a ratio scale – we cannot state a weight as –5 kg – whereas measures of temperature or direction by compass bearing or currency are in interval form. In the last three cases arithmetic can be performed on either side of the zero point. In practical terms this does not often affect the way in which data are actually handled, although there are certain scales, like the compass rose in which 359° and 1° are almost identical, which need special treatment. Both ratio and interval data can be given the full range of arithmetic treatment. Only when we have ratio or interval data like those of the Expenditure column of Table 3:2 are we able to calculate the mean or average as a foundation for more systematic arithmetic analysis. This increased power is important. Interval and ratio data types are sometimes grouped together as 'Quantity data', set against a 'Category' type which consists of nominal and ordinal data.

A couple of very important points should be made here. The discussion so far has been set in the framework of conventional decimal arithmetic. However, set theory offers a quite powerful but different way of handling nominal and ordinal data. Anticipating what will be said in Chapter 5, it also has to be said that despite their weakness in arithmetic terms, nominal and ordinal data can be subjected to statistical tests capable of producing significant results.

One final important point should be made. Numerical data should be handled at a level of precision suitable for the context. At one extreme, arithmetic performed on truncated or rounded numbers will produce inaccurate results. At the other, numbers displayed with an apparently high level of precision can be misleading. In the early 1970s one of the older universities carried out a programme of metrication which involved the replacement of signs on the premises. This was done with more enthusiasm than sense, so that for example the headroom restriction on one car park warning sign was converted from 7' 7" to 2.7644 m providing a sub-millimetric level of precision quite inappropriate to users' needs. For every plainly ludicrous example like this, however, there are many more in which judgement has to be used. It is reasonable to say that three of the seven categories in Table 3:2 above involved expenditure of more than £700,000. It is quite legitimate to express this as 42.86 per cent, but to do so is to give an unwarranted impression that the figure is derived from a much larger population.

Evaluating data

No matter what type our data belong to, we have to know the degree of confidence with which they can be used. Some of the problems of organising the process of data generation have already been discussed in Chapter 2 in the context of project design. There are other issues which relate to the generation process itself. In physical geography, data generation is often carried out instrumentally, within well-defined limits of instrumental and operator error. Human geographers less often have the chance to gather data instrumentally, and this makes it all the more important to be aware of the problems involved in the process. This is particularly true of any kind of interview or questionnaire work. Even the best-intentioned of respondents can provide incorrect answers because of misunderstanding or lack of knowledge. Particular cases, like the discrepancy between voting intentions as summarised by opinion polls on the eve of the 1992 British General Election and the voting realities of the election itself, have given us salutary warnings about the danger of people giving interviewers what they believe are the expected responses, even if these belie their real beliefs and intentions.

To be acceptable, data need to be both reliable and valid, and it is important to distinguish between these two characteristics. Reliability has been a major concern in the social sciences, and is quite easily defined technically in the context. A useful starting point is de Vaus's statement that a reliable measurement is 'one where we obtain the same result on repeated occasions' (de Vaus 1996: 54). Reliability means that the same observer asking the same question at different times, or different observers asking the same question at the same time, should consistently get equivalent results. Cross-checking individual data in this way is important in evaluating the data generation method, but in real life data will be collected in quantities that make it impossible for every data item to be checked. The test–retest method sometimes used by social scientists, which involves repeating questions to the same people at intervals of a few weeks and comparing the results, is not really feasible beyond the pilot study stage. What is important here is that the method itself is reliable, producing data that consistently fall within a given range of expectation. In this sense reliability is the degree of consistency with which observers put instances into the same categories. This reliability is akin to the replicability of the deductive scientific method, and for this reason has not been accepted as relevant by the school of social science that rejects scientific 'positivism'. Its members argue that the social world is constantly changing, and thus can never offer replicated findings. However, as Silverman has pointed out, this position is strictly tenable only if we assume that the world is in a state of infinite flux. Since systematic research would be pointless in such a case, anyone undertaking research implicitly accepts a situation in which some degree of replication is possible (Silverman 1993: 145–146).

Reliability is a necessity, and its practical significance hardly needs to be emphasised. The ease with which we can achieve it will depend on the type of data. On the whole it is easier to assess the reliability of numeric data generation. In a good many cases existing practice provides targets against which error levels in counting and measurement procedures can be assessed. Failing that, carrying out a pilot study and monitoring procedure during the generation process can establish a level of local control. Published data like census tables have the disadvantages of having been pre-processed, so that the original mode of data generation cannot be checked. However they will often have been rigorously checked before publication, and major data sources like national population censuses also generate their own literature that allows the user to evaluate the source before use (see for example Dale and Marsh 1993). The British Office of Population Censuses and Surveys (OPCS) is required to protect the confidentiality of individuals and households when it makes tables of census data available. This poses a problem in some cases where there are so few entries in a very small area that a determined user might be able to match them to individual households. The problem is solved in a number of ways. Sometimes tables are simply not released if their population size falls below a set level, but in other cases the individuality of cases is masked by adding 'random noise' to the values of cells (Dale and Marsh 1993: 77).

This raises two issues that will often have to be tackled. Although the deliberate changing of data values described above might be unusual, it is always possible that pre-processed data sets of this type will contain errors, and these errors might be related to the scale of the data generation unit. Missing values are also very important. It is not safe to assume that gaps in a data series can be treated as random accidents. When we come to analyse the results of an interview programme we cannot ignore the possibility that the interviewees who failed to answer a particular question, and indeed those people who refused to be interviewed, might actually have formed a population consistently different from the group which yielded the survey results.

In any kind of questionnaire or structured interview work, it is vital that the field observers are trained to offer questions to their interviewees in a consistent way, and to interpret the responses equally consistently. We will look at this in more detail in Chapter 4. Without this discipline, which is just as necessary if the questionnaire is administered by one person as a student exercise, the results cannot have any collective value. However, we should be careful not to regard reliability as an end in itself. It is quite possible for results to be utterly reliable and at the same time wrong or meaningless. A question might be designed and posed in such a way that respondents consistently answer it in a way that misrepresents their real beliefs or judgements. The results are 'reliable', but being consistently wrong is hardly the kind of reliability we seek. Nor is there any guarantee that a question which yields reliable results is worth asking. In fact we might expect that in any series of questions reliability will become progressively less easy to maintain as we move from the mundane and factual to the contentious and abstract.

If we move on from the question of reliability to that of validity, the issues are rather wider. For data to be valid they must not only be reliable but appropriate for the task in hand. No matter how confident we are that the data have been collected in the most reliable way possible, they only have value if they are appropriate for the intended context. If we look at the problem in this way, it has to be accepted that the concept of validity has to be attached not to the data themselves but to the way in which we use them. We therefore come back to the issues of appropriateness and relevance we examined in the first paragraph of this chapter. In fact we have to broaden our view to include issues raised earlier about project design (Chapter 2), issues we will consider later in the context of questionnaire design and statistics (Chapters 4 and 5), and indeed issues of methodology and philosophy that are not really within our brief at all. However it is worth thinking about validity here, since this is one of the important areas in which we can see data not simply as units of information but in a context where their collective significance can be assessed.

Social scientists have tackled this problem of validity by recognising three main types of validation, clearly summarised by de Vaus (1996: 56–57).

Criterion validity tests a new measure against some established criterion or benchmark. In other words, if existing research has established a measure which is accepted as providing a good assessment of a concept, a new measure can be devised and evaluated by comparing a population's answers to different questions based on the established and new measures. Good agreement indicates that the new measure is acceptable. However there are problems. Even if an acceptable benchmark is available, which will not always be the case, how was its reliability established in the first place? This approach offers no absolute measures. Furthermore, if the new measure is being developed because the existing ones are thought to be inadequate, they obviously cannot be used as criteria for its testing in this way.

Content validity is based on the process of verifying that the test as applied will cover all the fields thought appropriate. Thus if shoppers are to be surveyed to explore the way they get access to shopping centres, the questions might include one that simply finds whether or not individuals have access to a private car. This may be quite adequate in context. However if the survey aims to look in more detail at shopping by car, the concept of private car access has to be expanded to include issues like exclusive or shared use, co-ordination with other uses like the school run, disposable income, and so on. Content validity is not a powerful method of validation. It is not much more than a means of ensuring that the questions asked match the purpose for which they are being set.

Construct validity evaluates a measure by assessing it against theoretical expectations. Returning to the retail example, if we want to establish an index score of access to a car for shopping, and we expect that the value of the index rises with social class, the questionnaire could be designed to give us data that will allow us to test the relationship against a previously stated hypothesis. If the index scores do rise with class, the index has construct validity. This procedure shares the problems of other forms of hypothesis testing.

We should recognise that issues of reliability and validity, important as they may be, are simply means of ensuring that we are using the right data for our purposes. How we then go on to use those data is another and more important matter.

Summary

The success of any exercise depends among other things on your awareness of the quality of the data on which it relies. Being able to take full advantage of the strengths of your data without being misled by its weaknesses is one of the foundations of good research. Data should always be evaluated critically and not given a level of credence that they do not deserve. To paraphrase a memorable phrase used by Hugh Prince in describing the special problems posed by historical documents, data may be faithful servants but they are inadequate masters.[8]

Discussion questions and tasks

1 Test yourself on your ability to match different data sets to data types.

2 Take any United Nations Demographic or Statistical Yearbook and look at the footnotes to the tables. What does this exercise tell you about the data?

3 Does the difference between reliability and validity matter?

Further reading

See also

Research and project design, Chapter 2
Questionnaires and interviews, Chapter 4
Statistics, Chapters 5 and 6

General further reading

The problems of data looked at above are discussed in more detail in:

O'Brien, L. (1992) *Introduction to Quantitative Geography*, London: Routledge.

de Vaus, D.A. (1996) *Surveys in Social Research*, London: UCL Press, 4th edition.

Silverman, D. (1993) *Interpreting Qualitative Data*, London: Sage.

Notes

1   The word 'data' is a Latin plural. Like 'media', it is so frequently treated as a singular noun that it is probably only a matter of time before the 'correct' form fades into history.

2   They are also linguistically suspect. A datum is literally 'something given', so it cannot be constructed. However both 'fact' and 'fiction' once meant 'made-up things', so we should not let ourselves be bound too much by linguistic origins!

3   See the Supplementary Note to Harvey, D. (1963) 'Locational Change in the Kentish Hop Industry and the Analysis of Land Use Patterns', *Transactions of the Institute of British Geographers* 33: 123–144, reprinted in Baker, A.R.H., Hamshere, J.D., and Langton, J. (eds) (1970) *Geographical Interpretations of Historical Sources*, Newton Abbott: David & Charles, 243–265.

4   This review by Gherardi, S. and Turner, B.A. in 1987 is difficult to find, but an entertaining quote from it can be found in Miles and Huberman (1994: 49n).

5   In this particular case the form of words used in the primary data source to describe categories of spending would probably differ from city to city. The researcher will have to decide how they were assigned.

6   Derived from Yelling, J.A. (1986) *Slums and Slum Clearance in Victorian London*, London: Allen & Unwin.

7   Although some statistical tests are based on the summing of ranks (see Chapter 5).

8   See Prince, H.C. (1971) 'Real, Imagined, and Abstract Worlds of the Past', *Progress in Geography* 3: 23.

4 Questionnaires, interviews, and allied techniques: getting information from people

Research by human geographers often involves eliciting information from people. This chapter covers:

- Design frameworks and time-scales
- Interviewing key individuals
- Questionnaires and structured interviews – an introduction
- Sampling frames and sample sizes
- Designing the questions
- Administering the questionnaire
- Processing the results
- Working in the qualitative tradition
- Interviews
- Focus groups
- Participant observation
- Other qualitative techniques

In this chapter we are going to concentrate on three key areas likely to be important in student research, interviews with key 'actors' in processes, quantitative research in the form of questionnaires, and the use of qualitative techniques to investigate individuals' social attitudes and behaviour.

People as individuals are among the most important sources of information available to human geographers. The information they have to offer will of course take a wide range of forms. Some people can be classified as 'experts', in the sense that they are in a position to offer authoritative informed views of specialised fields. We can expect that the information they provide will be clear, consistent, and accurate, although it has to be said that we will sometimes be disappointed. However, everyone (including experts) has personal attitudes and opinions about a wide range of topics, and we can anticipate that these will be much less consistent and objective. Individuals also have their own characteristic ways of behaving spatially. To take a handful of examples, they use their leisure time in different ways, they organise their journeys to avoid perceived hazards, and they allocate their incomes according to perceived needs and objectives.

Expert opinion tends to be readily accessible. Experts write books, articles, and reports. They contribute to conferences, give evidence to inquiries, appear in television discussion programmes, and some of them now support their own web pages on the Internet. On the other hand they do not give us a direct entry into the rich field of attitude, opinion, and behaviour that is so important in the study of the ways that societies operate in geographical space. Information of this type can usually be obtained only by asking individuals questions, inviting them to talk, or using ethnographic methods to explore their worlds.

The techniques that allow us to explore this whole field of information are the basis of this chapter. A running theme will be the use of different techniques to explore the impact of a major retail development on shoppers' attitudes and behaviour. Retailing provides us with a good test bed, providing an arena in which major decisions are increasingly made by a small number of actors, but at the same time one in which the participation of almost the whole population in the shopping process provides a continuous test of the validity and appropriateness of these decisions in different contexts.

In Chapter 2 we made the distinction between intensive and extensive research. As we saw, extensive research treats individuals as members of populations, and is concerned with identifying regularities in their behaviour. On the other hand intensive research focuses on the behaviour and attitudes of individuals in their own right, pursuing motivation and causality directly rather than deriving them from the analysis of general patterns. Fundamental as this distinction is, it does not require that we must always choose one pathway at the expense of the other. There is no reason why they should not be used in conjunction, and this chapter intends to show how different approaches can be used to reinforce each other in the pursuit of a research aim.

A point common to all research involving individual responses is that we should not expect objectivity from our respondents. Generally speaking, participants in exercises such as questionnaire surveys or focus groups will be ordinary people without analytic training. They will have their share of the misconceptions, prejudices, and received opinions that burden us all. We have to accept this. Our aim should be to obtain as true a record as possible of what people think about particular topics or do in particular contexts. How we then interpret it depends on the aims of our study. Experts pose different problems. If you are interviewing key actors in (for example) the debate about planning permission for a shopping centre, you have to be aware that expertise and impartiality do not go hand in hand. Consciously or unconsciously, your respondents will tend to filter the information they provide on the basis of their own viewpoints and their organisations' policies. If you are not an expert in the field yourself, it may be very difficult to tell when filtering is happening.

A final important general point is that we cannot and should not expect total impartiality from ourselves. The debate about the genuineness of scientific objectivity has been running for many years, and the concept of the objective researcher is particularly difficult to sustain in many fields of human geography. In other words we have to recognise and take into account our own involvement in the processes we are studying. Ethnographic techniques, in particular, put researchers in situations where academic distance cannot be maintained. A familiar defence of the ethnographic approach – and a reasonable one – is that researchers whose methods involve them in interaction with the people they are studying are constantly kept aware of their own subjectivity. Ethnographers think that workers using positivist extensive techniques, in contrast, run the risk of over-estimating their own detachment (Cook and Crang 1995: 11). All human geographers have to strike a balance. On one hand we have to recognise that our own cultural loading will affect the way that we do research. On the other, we should not let the fact that we are exploring a subjective world of individual attitudes encourage us to abandon any kind of rigour in our exploration of it.

The overall aims and planning of any research exercise eliciting responses from people must be seen in terms of research and project design, fields we considered in Chapter 2. The data generation techniques examined in this chapter can only produce valid results if they are used appropriately within a carefully defined research design.

Design frameworks and time-scales

For the reasons already mentioned, we have to be confident in advance that we know why we are carrying out the exercise and how it will be organised. Our purpose will determine which population we are going to work with and we also have to consider the important issue of the time-scale in which we plan to work.

The commonest form of data generation, particularly with the limited time available at student level, looks at an individual episode and does not take time-scale into account. Although it may be necessary to extend data collection over a number of days or even weeks, this is more likely to be the result of limited resources than deliberate design. The term 'cross-section' is sometimes used to describe this form. As examples we might think of an ethnographic study of the significance of shopping in individuals' lives at a particular time, or a retail questionnaire survey in which we explore the impact of a new development on shoppers' choices between different centres on the basis of age or income or mobility.

Change over time is implicit in many cross-sectional studies. In the examples above we would benefit from the ability to carry out later 'follow-on' surveys, although these are almost always beyond the scope of student projects. If analysis of our shopper questionnaire shows that there is a significant change in preferences with age, we might feel justified in claiming that the younger groups in our survey will eventually take the same course. In other words a similar exercise carried out ten, twenty, or even thirty years later would be expected to show the current younger age bands changing their attitudes and expectations over time. However, this is not a safe assumption. We may find that there is an association between age and behaviour, but we cannot establish that it is a direct causal relationship, since we know that the young shoppers of the 1990s have grown up in a different social environment from that of their older counterparts. Because our ethnographic analysis allows us to explore attitudes in much greater depth, it should give us a much more confident grasp of the real relationships between age and attitudes. Even in this case, though, we are in no position to anticipate the ways in which further social changes might affect peoples' behaviour over the next twenty years.

One solution to this problem is to carry out what is sometimes called a 'before-and-after' study. A typical before-and-after research programme will be planned in anticipation of a major change in a variable affecting individual behaviour or attitudes. For instance we might be interested in the results of a change in the law affecting housing tenure, the completion of a by-pass scheme, or a well-publicised customer service campaign providing free transport in order to woo non-car-owning shoppers into using a new out-of-town shopping centre. The distinguishing feature in this kind of research is that the individuals questioned or interviewed at the 'before' date also form the group questioned at the 'after' date. In principle at least a study of this kind is a powerful tool, since concentrating on known individuals means that we can eliminate random variation in the population. Using qualitative techniques to analyse changes in individuals' attitudes and social relationships over time is an important step towards understanding causality. In a world without resource constraints, a long-term study integrating quantitative and qualitative approaches would offer us the best way of analysing change.

Returning to our retail questionnaire example, a before-and-after study might be able to demonstrate that immobile shoppers have changed their preferences in apparent response to a free transport campaign but we have to allow for the possibility that over the period in question the store chain has managed to improve its market share globally by a similar proportion. In such a case the non-car-owners would simply be responding in a non-specific way to a factor or factors for which we have not tested. For this reason it is normal to set up a control group in a before-and-after questionnaire survey. This control will normally be designed to exclude individuals who might have been selected for our initial group. In this case the control would include car-owners and others who were not likely to be affected by the campaign. If the control group does *not* show a change in performance, we can be much more confident about our ability to interpret our target group's behaviour.[1]

The logical development of the before-and-after approach is the longitudinal survey, in which individual respondents are revisited at intervals over a long time-span. In an ideal world we could set up a framework in which the same shoppers were contacted at regular intervals, perhaps annually over a period of ten years. We would thus be able to provide interim and final reports on the way that these individuals had changed. It does not take much thought to realise that longitudinal surveys are formidably difficult to manage. The number of respondents inevitably will have to be much larger than we would regard as satisfactory for a single survey, since we must allow for wastage and drop-outs over a long period. If we were working quantitatively, it would probably be necessary to increase the sample to include a non-age-specific group as a control for each date. There also has to be a stable working base and a secure source of funding to cover the costs of administering a research programme for such a long time.

For these reasons no student is likely to have much involvement with a longitudinal survey except as part of the sample. Even a simple before-and-after survey might demand time and commitment beyond the normal range of student projects. However it is important to recognise that the possibility exists. In some cases it might even be possible to turn to good use the tendency for undergraduate research to extend over longer periods than would be needed for a well-funded survey with adequate field staff.

Interviewing key individuals

Student research often involves contact with key individuals. These are the 'experts' we looked at earlier, although their significance to us usually lies in their position within organisations rather than directly in their expertise. Typically they are decision-makers in companies, governmental agencies, or pressure groups. We will usually want to use them as sources of information about the behaviour and policies of their organisation in particular context. If we are looking at a new retail development, for example, the key actors are likely to be councillors, and local authority officers such as members of the planning department, representatives of the development company and major stores, and perhaps members of pressure or protest groups. How you identify the actors and make contact with them will depend on the nature of your research.

These people are special in a number of ways. First of all, your study is likely to make use of very few of them and each one will require a different approach. A second key point is that although you will know in advance the general areas you want the interview to cover, you have to be sensitive to events on the day. A respondent might unexpectedly veto one line of questioning, but be much more forthcoming than expected in another field. For both these reasons the approach must be qualitative rather than quantitative. There is no scope for formal questionnaires and numerical processing. A great deal depends on your ability to focus the dialogue on the areas that concern you.

Most of the general issues relating to interviews with key actors are essentially matters of common sense.

Many of the people you will be dealing with will be genuinely busy. They will also be used to being in control, and since you are asking favours of them you will have to accept this. It may be difficult to arrange an interview. It will probably be necessary to negotiate with personal assistants or other gatekeepers, and gracefully accept a short session at a time and location that suit the interviewee. If you break the appointment you are unlikely to get a second chance. It is also important to prepare as much ahead of the interview as possible. In this context it is not unlikely that the interviewee will try to control the agenda, and unless you are well briefed you will probably not be aware of the direction in which it is being steered. The interviewee is likely to take more seriously a student who shows an awareness of the issues that matter and an ability to speak the specialised language of the field, although too aggressive an attempt to show your recently acquired knowledge may be counter-productive.

The interview itself needs some care. You should explain clearly at the start how you plan to record the information you get, and get agreement for it. A hidden tape recorder is not a good idea! Start the discussion with safe establishing questions, and save the more sensitive issues for later. It is difficult to offer general advice on how to handle an interviewee who evades these. Pursuing the points directly may work, but if it does not you may have to change topics and try to pick up information obliquely. Don't indulge yourself by getting involved in an argument. This will probably bring the interview to a speedy end and close the door on other interviewers in the future. Members of special interest groups are generally quite approachable. They are after all representatives of bodies which exist to communicate a particular point of view. However this enthusiasm in itself makes them more likely to be obsessive, and you may have to steer them gently away from their favourite topics.

Interviews of this kind should not be undertaken unless they are strictly necessary. There is a limit to the number of interviews that even the best-natured expert can accommodate, and badly managed discussions will reduce the level of goodwill sharply. Some academics are worried, and not always without reason, that unpolished student interviews might alienate sensitive populations and make it all the harder to carry on higher-level research. For example there are researchers working with the management of small businesses, a difficult group at the best of times, who will admit off the record that they have discouraged their students from carrying out interviews.

Questionnaires and structured interviews – an introduction

Questionnaires and structured interviews provide a widely used way of putting the information that people provide into a quantifiable framework. Inevitably they do so at a cost, as we saw earlier in comparing the outcome of extensive and intensive research approaches. However they have an important role and are commonly used in student research.

Questionnaires and structured interviews are means to the same end, i.e. obtaining data in controlled conditions, to enable us to make acceptable inferences about populations[2] from a sample of individuals. In both cases we obtain individuals' responses to a range of pre-defined questions, and the distinction between the two forms is not at all clear-cut. Strictly speaking we should perhaps use the term 'questionnaire' to include only self-administered and postal examples where there is no personal contact with a field interviewer, but the term is often used to cover cases where structured interviews are carried out face-to-face or by telephone. We will take this more relaxed view in this chapter, but the point is still an important one. If you are conducting a structured interview face-to-face you should always be alert to the possibility that personal interaction between the respondent and an interviewer might affect the outcome, and the interviewer might unwittingly filter or interpret the responses as they are entered on the coding form.

How do we know that the inferences we get from questionnaires are acceptable? At the simplest we do so by ensuring that our questionnaire obtains information from appropriate people in appropriate numbers. In other words the sample should draw its responses from a selection of individuals who can be taken as collectively representing the population to which they belong, and it should also include a large enough number of individuals to mean that our results can be accepted statistically as presenting a true picture of the state for the population as a whole. A sample which meets these criteria is said to be 'representative' of the population. At the risk of stating the obvious, the data should also be useful. In other words the information actually obtained must not only be an accurate reflection of the individual cases but should be capable of analysis that will yield answers to questions that matter. We also have to be able to work out how large a sample is needed. We will think more about sample size at a later stage.

When should we use questionnaires? As in the case of interviews with key actors, we should do so only when they are appropriate and there is no other way of generating data. There are good grounds for being sparing in the use of the technique, and quite a strong case can be argued for avoiding the use of questionnaires except when absolutely necessary. With very few exceptions,[3] questionnaire surveys are based on information volunteered by individuals as a matter of goodwill. The fund of goodwill is limited. The growth of survey research in all its forms means that some populations – the inhabitants of New Towns and some metropolitan housing areas come to mind – have been deluged with questions to the point that 'questionnaire fatigue' has set in. It will take excellent presentation and a persuasive touch to get an adequate response from a jaded population like this. Apart from over-exposure contributing to the loss of any sense of novelty, there is a long-standing mistrust of questioners who might turn out to be 'somebody from the Council' looking for information for some sinister official purpose. More recently people have also learned to suspect that a seemingly objective questionnaire might turn out to be a commercial marketing exercise. Although the population as a whole has a larger collective reservoir of goodwill than the relatively small number of key actors, it is still finite, and a really badly administered one might leave some participants unwilling ever to go through the experience again.

Another important aspect of questionnaire research is that it usually works on a one-pass basis. There will be only one chance to tackle any respondent.[4] If the project design calls for the sample to be taken at a particular time, any later modifications may be methodologically invalid. Even if this is not the case some sampling methods provide no means of locating respondents once the questionnaire has been administered. Where address-based sampling has been used it is possible for the researcher to contact the respondent again, but repeat calls to tidy up the answers are unlikely to be sympathetically received. In any case questionnaires and interviews are expensive in time and money, and it may be unduly expensive to try to rectify any mistakes or omissions. The questions must be right first time, and any weakness in question design or the method of collecting responses will be punished cruelly. The most reliable way of avoiding this problem is to undertake a pilot study, testing the operation of the questionnaire on a small population and making whatever changes seem to be necessary. We will look at pilot studies in more detail later.

Sampling sizes and sample frames

Having established what population we are going to sample for our questionnaire, we have to ensure that our sample is representative. As we saw earlier there are two main criteria for this. The selection of cases from the population must be done representatively, without any deliberate or unwitting bias in the process. We have to devise a sampling frame – a list of the members of the population – and draw cases from it. We will look at sampling frames shortly. First, though, we have the issue of sample size. The sample has to be big enough for us to be confident that any analysis made of it can be convincingly related to the population it represents.

Sample size is a difficult issue, and calculating the optimal sample often involves a difficult balance between our desire for accuracy and constraints of time and money. A sample can be defined in terms of 'sampling fraction', the proportion of the total population it includes. A sample of 50 from a population of 450 has a sampling fraction of 450/50 or nine. However for reasons we shall look at shortly, figures based on population size may not be very helpful to us. Statistical techniques can be used to determine how large our sample needs to be. If we are able to assume that our population is normally distributed (see Chapter 5) we can use a formula based on levels of sampling error. Suppose that a planned questionnaire for non-car-using shoppers contains a question about their readiness to shop more often at the store if transport were to be improved. We may decide that we want the sampling error for this question to be 2 per cent. This means that we want to be confident that our sample 'errs' no more than 2 per cent from the response we would have got by questioning the whole population. Thus if our sample yields a 38 per cent positive response to the question, a sampling error of 2 per cent means we are able to assume that the figure for the total population will lie between 36 and 40 per cent (i.e. 38 ± 2 per cent). This is an estimate of probability rather than a precise statement, and we will also have to set a 'confidence level' – normally 95 or 99 per cent – at which we can accept that this range is correct (see Chapter 5).

If we fit our figures into an appropriate formula[5] we can show that at a 95 per cent confidence level we need a sample of 2,262 to achieve a ± 2 per cent sampling error. The required sample size falls rapidly if we relax our requirement. If we opt for a ± 5 per cent range of acceptable error we need only 362 cases, and at ± 10 per cent the sample needs only to be 90. Naturally explanatory power is much lower. At the lowest of these levels we can only say that we are 95 per cent confident that something between 28 and 48 per cent of the population as a whole would give a positive answer to the readiness-to-shop question.

Two important points emerge here. The first is that the estimate of sample size is derived from confidence levels and our acceptance of levels of sampling error, and is not dependent on population size. The second point, illustrated by the changing ranges of acceptable error, is that the required sample size falls strikingly as we reduce the stringency of our demands. Few student projects are likely to be able to generate the samples of 2,000 cases or more needed for a really low range of error, but a sample of 100 or less is quite practicable and can still provide acceptable results as long as we are aware of its limitations.

The choice of sample size will always be a compromise, but if in any doubt a larger sample will always be better. The sampling error estimates described above are rather generous, since they assume an error-free sampling operation and complete response. Non-response means that the actual sample will virtually always be smaller than the expected sample size. The type of sample may have to be taken into account. If the exercise is likely to yield a large number of sub-groups of information and statistical tests are to be used, the sample size may have to be increased so that we generate enough cases in each group to fit their needs. If data are likely to be recorded in nominal or ordinal form, it might also be necessary to increase sample size to achieve significant results with the less powerful parametric tests that will be available (see Chapters 5 and 6).

Having worked out how many people we need in our sample, how do we find them? In other words how do we build a sampling frame? This question will inevitably be defined by the purpose of our survey. Usually the target population itself will form only part of the population as a whole, and sometimes a very small part. Locating a sample 'blind' could therefore be extremely difficult. Suppose that we wanted to survey the environmental attitudes of beekeepers. There are about 30,000 amateur beekeepers in England in a population of about 50 million, a proportion of 0.06 per cent or approximately 1 in 1,700 of the population. Clearly a randomly generated survey of individuals in the population at large would be very wasteful, with hundreds of thousands of questionnaires needed to locate even a modest sample. However, about 12,000 of the beekeepers belong to local associations within a national framework, and an approach to a local society might be a fruitful way of constructing a sampling frame.[6] In all cases where the target group is associated with membership of a trade organisation, special interest group or pressure group, there will probably be a membership list with national or local addresses and contact numbers. If the organisation is willing to release the list it may also give its official approval to the survey, a major benefit if some members are reluctant to co-operate.

Official records relating to education, health, employment, taxation, car ownership, crime, and other fields all allow individuals and households to be targeted on particular criteria. For reasons that need not be stressed, many of these will not be available to student researchers. Some of them, like medical, fiscal, and criminal records, are obviously sensitive and are unlikely to be available even in 'anonymised' form. It is also possible that the way in which the data have been gathered make it difficult to draw a representative sample.

Where the target population does not have such a well-defined identity, more general sources have to be used. Telephone surveying has a long history. The telephone directory is a convenient source, particularly now that it is available in Britain and elsewhere on CD-ROM. It is most helpful for surveys of official bodies and businesses, since few organisations do not have a telephone and not many businesses have any interest in concealing their contact number. Residential addresses are more problematic. Roughly 15 per cent of households have no access to a telephone, and a larger proportion (25 per cent of numbers at the start of the 1990s and certainly much more now) are ex-directory. A directory search will reach only about half of the population. Sampling is most easily done by drawing randomly from the directory, but for more advanced work a technique called 'random digit dialling' allows the user to generate random telephone numbers within geographically-defined area codes. Some of these numbers may be unallocated, but this method also potentially brings in ex-directory numbers (Oppenheim 1992: 97–99).

In Britain the electoral rolls provide a reasonably comprehensive and accessible list of names and addresses. They are always a little out of date because of changes of address (the same is true of the published telephone directories), but a bigger limitation in some contexts is that they include only adults of voting age. An additional problem stems from resistance to the Poll Tax in the 1980s. Anecdotally at least a sizeable number of people are said to have chosen not to register in the hope of avoiding detection as potential tax payers, although by the mid-1990s the size of the electoral roll had grown enough to suggest that most of them had returned to the fold.

A very traditional geographical way of sampling addresses is of course to use large-scale maps as a sampling frame. Although there are obviously problems with households 'stacked' vertically above each other in blocks of flats and particularly estates of tower blocks, the map provides a useful framework in the public domain for geographically structured sampling of the type described later. Piecemeal revision means that published maps tend to be out-of-date. Unlike the cases of telephone or electoral roll sampling, however, a researcher with a small area to sample can take the map out and update it in the field before the sampling takes place.

There is no perfect sampling frame. A question we must always ask is whether or not any deficiencies we know about can be treated as random or systematic. In the case of the beekeepers, rather more than half of the enthusiasts in England do not belong to a society. Perhaps there is no significant difference between joiners and non-joiners. However, if we were to carry out a survey drawn only from members, we should always bear in mind the possibility that the decision not to join may point to a more general difference of attitudes within the non-member group. This problem will confront us whenever we meet a self-selecting group like a special interest society. In the telephone case there may be no obvious common identity among those who choose to go ex-directory, but traditionally the relatively poor, the young, and those who change address regularly have tended to be prominent among those without access. Any telephone survey should bear this in mind. In the case of the electoral rolls, the anecdotal evidence about Poll Tax avoidance points to a temporary selective withdrawal of relatively young people in the lower income brackets living in casual or fluid household arrangements. In many ways this is also the group least likely to have telephone access.

It is always important to know the characteristics of our source of cases. The examples above show how particular sources might fail to represent the population as a whole. In some cases this might mean that a source cannot be used at all for a particular purpose.

So far we have considered how to determine our sample size and considered how to build a sampling frame. How do we actually select cases? One of the commonest techniques, and an important one if we want to carry out a statistical analysis, is to take a random sample. Random numbers for this purpose can be obtained from published tables or generated by computer. In a truly random sample the process is organised so that every member of the population has an equal and independent chance of selection. Independence in this case means that inclusion of one individual in the sample does not increase or decrease the chance of any other being included. We have to be careful about this. If we start with a population of (say) 450 and plan to take a sample of 50, when the first choice is made the probability that any one will be picked is 1/450 or 0.0022. If each case is withdrawn from the sample when it has been selected, by the time we have selected 49 the odds against one of the remainder being selected as the 50th case have shortened to 1/401 or 0.0025. Small as the difference might be, the rule about independence is still being broken. Strictly speaking every case should be drawn from the same total population, and in principle at least if the same case is selected more than once it should figure in the analysis as often as it is picked. This is perhaps not as commonly done as it should be.

If we are using a membership list or some other kind of population summary on paper as our sampling frame, we do not have to take space into account. However, there will be other cases where we are actually sampling distributions on a surface, and we have to be aware of the way that these distributions are organised. Human geographers may not be as likely to use simple random sampling as their physical counterparts. Human behaviour tends to be organised at particular points and along lines, rather than continuously distributed like soil moisture or ground cover. However simple random sampling is sometimes used. Figure 4:1a simulates a section of a spatial surface (a city) sampled randomly. If we want to interpret this pattern it is important to know how the sampling was done. The researcher may have used random numbers in pairs to define grid co-ordinates and then sampled the households nearest to them. If this was the case, any relationship between the distributions on the map

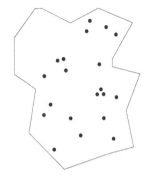

(a) Simple random sampling

(b) Systematic sampling

(c) Stratified random sampling
(Each sub-area is randomly sampled)

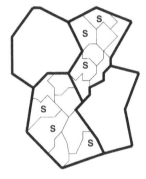

(d) Cluster area sampling
(Areas marked 'S' are sampled)

Figure 4:1 *Sampling frameworks*

and reality on the ground can only be coincidence. However, if the map shows the result of random selection from a list of all the city's households, the situation becomes more complicated. Do the little clusters in the sample reflect the high density of households in inner-city areas, which makes it more likely that some will be picked, or do they simply result from random selection? We might ask a similar question about the empty areas, particularly in the top left. Is this a zone in which scattered suburban villas in spacious grounds offer few potential cases, or has our sampling procedure simply failed to generate any instances here?

Systematic sampling replaces randomness with regularity, selecting individuals from the population on a structured basis. Sometimes it will be the more appropriate method. Taking every *n*th name from a list or sampling pedestrians by stopping every nth one is much easier than working from a random number table, but there are problems with this method too. The moment a systematic frame is adopted, it is inevitable that certain individuals will be included and others excluded, and this has significance. Strictly speaking it means that parametric statistical tests should not be used, since potential members of the sample do not all have an equal and independent chance of inclusion (see Chapter 5). It is also vital that the choice of frame is made sensibly. The biggest danger is that the sampling interval will coincide with some cycle in the sampled data. A well-known cautionary tale concerns a researcher (probably fictional) who selected the first and last entry on each page of an alphabetically ordered membership list. Although this was easy to administer and looked rational, its effect in terms of the list as a whole was to select pairs, some of whom were even related to each other! Geographers have to take account of spatial aspects of this problem. There is little danger of it with pedestrian sampling, but suppose that we are sampling residents of tower blocks to investigate their mobility problems. If we know that there are ten flats on each level we might be tempted to take every tenth number in order to cover each floor, but since flats are usually numbered consistently every one selected might occupy the same place on the floor plan. In some circumstances this might not matter, but if we happen to have selected flats particularly close to the lift, we may have built a bias into the results of any query about mobility.

Figure 4:1b simulates a systematic sample in the same urban area as Figure 4:1a. In this case our researcher has used grid intersections to locate the sample points, and has then interviewed the householder nearest each selected point. This method has limited merit. It is less likely that it will completely exclude any area, as can happen with simple random sampling. However the figure shows how the relationship between the area's boundary and a coarse grid like the one used here might make sampling round the perimeter rather haphazard. The more important problem is that this sampling scheme is only justifiable if the area is geographically uniform. Since cities are places with great internal variations in population density, composition, and behaviour, this kind of systematic frame would give us a very unrepresentative sample in this context.

A third approach is to use 'stratified' sampling. As we saw earlier, simple random sampling might exclude some areas purely by chance. We can ensure even coverage by using a stratified sample as shown in Figure 4:1c. Here we are using the administrative framework as the basis for our sampling method, and in the example shown every one of the thirty-five sub-areas will be separately sampled randomly at the same proportionate level. In this way consistent coverage is ensured. However, if we know that a characteristic that we want to study (perhaps the distribution of households lacking important amenities) is concentrated in some areas in particular, consistent coverage will not suit our needs. Here we might use disproportionate stratified sampling, in which we deliberately adjust the sampling proportion to ensure that a high proportion of our sample is drawn from these areas. Naturally this is only advisable if we are confident that we know how the distribution occurs. All forms of stratified sampling pose statistical problems. Although it can be argued that stratified random sampling allows parametric testing, the populations of the individual sub-areas used for sampling might not satisfy the assumptions of the normal distribution even if the overall population does (see Chapter 5).

Another method is cluster area sampling, which can be used where a hierarchical structure is available for sampling in spatial or non-spatial contexts. Figure 4:1d shows in a simplified way how it might be used in a city, although the process could equally well be carried out in a non-spatial hierarchy like an organisation's structure. First of all we randomly select members of the highest level of the hierarchy. This gives us a group of sample areas, and we then discard the others and move on to carry out another random sample of the next level within the selected areas. This process is carried on downwards as far as appropriate, leaving us with a final selection of small survey areas. Every individual case in these is then sampled. In our very simple example there are only two levels, and we have chosen two areas out of four at the first level, then three sub-areas out of 8–10 at the next. This method is most helpful if we are planning to carry out a household-based survey and want to reduce travel time and costs during the survey without prejudicing the structure of the sample.

It should be clear from the discussion so far that our choice of sampling method will depend on a number of factors, including the purpose of the survey, our prior knowledge of the population, and the methods of analysis we plan to use. Concise general discussions of sampling methods can be found in texts such as Oppenheim (1992) and de Vaus (1996). Interested readers will find a useful expansion of the techniques of spatial sampling in Ebdon (1985), and a study of the literature will show which methods have been found most suitable in a particular field.

Designing the questions

At this stage we can start thinking of ways of getting information from our sample. Good questionnaire design is not as easy as it might seem at first, and needs a lot of experience and thought.

Before a questionnaire is put into use, it should be carefully tested. We should be confident by the time we have got to this stage that we are quite clear about all the issues relating to the purpose of the survey, the population sampled, and the nature of the sampling process. Now we have to make sure that the questionnaire will meet our needs. It is vital that the questionnaire will fit the requirements of the project, and at the same time will work in the field to produce appropriate results. Some kind of pilot survey is essential. In large-scale professional survey work every aspect of the process will be piloted, from the design and appearance of the form to the problems that might emerge when the results are processed (see Oppenheim 1992: 45–64). Probably the most important part of the piloting process is that a preliminary form of the questionnaire is tested on a relatively small sample in order to make sure that no unforeseen difficulties are likely to turn up.[7] Thorough piloting can be very cost-effective in reducing wastage when the full questionnaire is run. It may also improve the academic quality of the work, by re-opening conceptual issues which had not really been considered before the pilot was run. Limited resources mean that a student questionnaire cannot normally be piloted in the same way, but a small pilot study might be run, and at the very least fellow students or friends can provide an informal first response.

What are the criteria for a good questionnaire? The way in which a questionnaire is worded and presented should be responsive to the characteristics of the target population, and if it is to be self-administered (as discussed in the next section) it must be completely self-explanatory. However the principles set out below can be taken as important in all cases.

Comprehensiveness and relevance are vital. Since there will be no second chance, every question that is important to the analysis must be included. On the other hand it would be counter-productive to throw in a large number of extra questions just in case the answers might be helpful. An efficient questionnaire that achieves the right balance is not easy to produce.

Brevity and the ability to sustain interest are also important. It is unrealistic to expect a respondent to spend more than a few minutes answering questions in normal circumstances,[8] and the response rate will drop off sharply if and when respondents find that they are expected to commit a lot of time. Since it is rare for a questionnaire to operate on any basis but goodwill, it should be designed as far as possible to be interesting, and introduced by a short explanation of its rationale using any reasonable way of pointing out the value that the response will have.

Sensitivity is also important. Some respondents will regard personal questions as intrusions on privacy. Requests for information about income, political allegiances, or religious beliefs are likely to have high refusal rates, and so are those which might incriminate respondents or put their behaviour in a bad light. Some respondents will be reluctant to give information on less obviously sensitive issues like age, marital status, or type of employment. As a general principle, questions on these topics should be included only where they are strictly necessary. If they have to be asked it is wise to introduce them on a voluntary basis at the end of the questionnaire. If it is necessary to get information about the respondent's home location ask for a postcode element (the first half may be sufficient) rather than the full address.

It hardly needs to be stressed that questions should be clear. There are several steps between the first conception of the questionnaire and the data analysis stage, and all of them have the potential for confusion. The initial idea has to be translated into a question, which then has to be posed to the respondent, who then formulates and communicates an answer. This answer in turn has to be recorded, and at a later stage the response is incorporated into the analysis. This gives us at least five points at which information has to be communicated, and things can go wrong at all of them. Here we are dealing with the central area where we have to minimise ambiguity and the potential for misunderstanding in the posing of questions and the recording of answers. We also have to look beyond our own survey and make sure that the information we get is compatible with other sources. For example, we should take care that information we collect about economic status or occupation can be related to the classifications normally used.

Even the simplest questions can be deceptive. If we return to the questionnaire about shoppers' mobility, our naive surveyor has included the very basic question – 'Do you have a car?' His first respondent thinks about her[9] response. Her husband has a van which he uses for his business, but his working hours are irregular and it is usually full of building materials, and he seldom lets her drive it or takes her to the shopping centre in it. She wonders whether the surveyor is concerned about the difference between cars and vans but thinks that the question probably hinges on its use for shopping. However, she also has an understandable reluctance to lose social status by letting it be thought that she belongs to a non-car-owning family, so she gives a 'yes' answer, which the field surveyor notes down as she trudges off to the bus.

This question is disastrously ambiguous, and should never have got past the pilot stage. It confuses accessibility with ownership, and fails to distinguish between ownership by the respondent or a partner or family member. It unwittingly narrows the range of transport by ambiguity over the status of vans, people carriers, and campers. By raising the issue of ownership, it becomes loaded with an element of social status and tempts respondents to make the most favourable interpretation. What he really meant was probably 'Do you have independent use of private transport for shopping trips?' Brevity is not necessarily clarity. Although questions should not be so full of sub-clauses that they resemble legal contracts, they must leave the respondent in no doubt about what is meant. If necessary a question as complex as the example we have used can be broken down into a sequence of individual elements.

Questions can be classified in a number of different ways. One important distinction is between closed and open-ended questions. All closed questions ask for an answer from a defined range. At the simplest they might ask for a precise item of information like age or the number of persons in the household, or an answer from a simple multiple choice like agree/disagree. A great many closed questions are classification questions asking the respondent to make a choice from a defined list, which will almost always include the escape-route 'Other' category. Open-ended questions invite the respondent to answer the question in whatever way might seem appropriate.

Some questions will naturally tend to need one of these approaches rather than the other. Where you have the choice of either form it is useful to think of their different strengths. Closed questions allow the surveyor more control over the process. Information from the respondents is collected in a highly structured form which should be easy to code and analyse. The strength of open questions is that they allow the respondent to provide information that the surveyor could not easily anticipate. However the answers to open questions may be relatively long and discursive. Recording this information accurately is a real problem, particularly if the questionnaire is being administered face-to-face or over the telephone. At the stage of coding for analysis, decisions might have to be made which would much better have been made with the respondent present. Coding and analysing open questions will always tend to pose problems that do not arise with closed questions.

Another fundamental distinction is between what Oppenheim calls factual and non-factual questions. Factual questions call for information about the individual that could in theory be obtained by external observation. An enormous team of private detectives monitoring shopping behaviour might provide more reliable information about shoppers' movements than a questionnaire, although it would be hugely expensive and probably illegal.[10] Non-factual questions are essentially those which explore respondents' attitudes, opinions, and beliefs. For obvious reasons these non-factual or attitude questions are often open-ended, but closed non-factual questions are also common, usually asking the individual to make assessments on graded scales covering ranges like 'Strongly disagree – Strongly agree' or 'Very bad – Excellent'. Very broadly, factual questions tend to be closed and non-factual questions are open-ended, but this is a tendency rather than a rule. Oppenheim suggests as a general principle that all questions should start life in open form, in the expectation that the pilot study will show which might be best converted to closed questions.

Filter questions are used to steer the interviewee into or past different sections of the questionnaire as appropriate. Careful use of filters can increase the flexibility and data gathering power of the questionnaire without requiring a great deal more from the respondent. However a questionnaire with a lot of filters can look daunting, and is best administered face-to-face or by telephone by a trained interviewer.

There are many pitfalls in the detail of question design, and the following list offers a warning about some of the most common.

- Avoid long and complex questions. Every question should be as short as practically possible, and should not address more than one idea.
- Avoid questions where the answer leaves an ambiguity. If the question 'Do you regularly travel by train or bus?' is answered by 'Yes', it means that the respondent uses buses, trains, or both. The surveyor cannot tell which.
- Avoid jargon or polysyllables except where the context demands them. In all cases the language used should be clear and appropriate for the target audience.
- Avoid leading questions of the 'When did you stop beating your wife?' variety.
- Avoid double negatives. A question like 'Subsidies should not be used to support bus companies. Agree/disagree?' is unduly taxing. Respondents who favour subsidies have to disagree with a negative.
- Do not strait-jacket the choice. 'Others', 'Don't know' and 'Not applicable' options should be used where appropriate

Much more comprehensive guides to question design will be found in the texts by Oppenheim (1992: 119–149) and de Vaus (1996: 80–105). An important point is that questions should be seen not just as individuals but as a group. The questionnaire as a whole should start with interesting questions. Interesting and open questions should be strategically placed throughout the series to retain but not over-tax attention.

Administering the questionnaire

All questionnaires have to be administered. That is to say, the piloted forms have to be introduced to the sample population, filled in, collected, and collated. There are two important points that cover all questionnaires. First it is important to get all permissions that might be needed. Second, the surveyor should make it clear to the potential respondents what degree of confidentiality is being offered, and should not deviate from it.

There are several ways of administering questionnaires. Often the questions will be asked directly by a fieldworker. In student-defined work this will usually be the originator of the questionnaire. Face-to-face contacts cover the range from brief pedestrian surveys to long formal interviews, and also include telephone interviews. In all cases the surveyor records the information and interprets the questionnaire if necessary, although if this has to be done frequently the pilot stage has not done its job.

The field surveyor is very important at this stage. He or she has to make first contact with the interviewee and establish sufficient rapport to ensure that the forms are completed, while at the same time ensuring that they are filled in consistently and accurately. Commercial survey organisations have traditionally preferred women in this role, not least because they are perceived as being warmer and less threatening than men. Male students administering their own questionnaires might have to work harder to achieve a suitable response rate. Racism can also be a problem. Some black male students administering shopper surveys with which this author has been associated have been disconcerted to find how poor a response rate they experienced in relation to other surveyors. The problem has a geographical dimension. Household surveys in some inner city estates can have particularly disappointing refusal or non-response rates.[11]

Self-administered questionnaires include postal surveys, those distributed to employees with the co-operation of their management, and others made available for public collection. An example of the last category might be a questionnaire made available to visitors to an archaeological or historic site with the agreement of the site supervisor. Self-administered questionnaires are generally more problematic. Since there is no field surveyor to advise and interpret, the questionnaire must leave no doubt in the respondent's mind about what is required. Response rates tend to be low, often below 50 per cent. Part of this is probably due to the lack of a direct incentive in the form of a clipboard-wielding surveyor, but there may also be elements of doubt about the purpose and confidentiality of the survey. An informative covering letter or introductory paragraph will help to some extent.

How can response rates be improved? In general self-administered questionnaires seem to work better if they offer a clear indication of the reason for the survey and the choice of the individual. Where members of an organisation are being surveyed, a letter of sponsorship or support from the organisation will be helpful, and it might also be useful if the organisation can give its members advance notice of the survey. The design of the questionnaire itself is particularly important, and it is vital here that the first few questions attract the respondent's interest. A rather obvious point, but one that is not always taken into account, is that where the response is to be returned in the post inclusion of a clean stamped return envelope will improve the response rate.

Working in the qualitative tradition

Questionnaires are effective research tools, but they have major limitations based on their highly structured mode of operation. They must be based on sizeable samples if they are to provide acceptable results, and the questions they contain have to be highly structured to standardise the range of response. In this sense the questionnaire technique is inefficient, processing a large number of respondents but getting relatively little information from any one.

A large number of geographers now prefer to use a range of qualitative techniques to investigate peoples' attitudes and behaviour. The richness of information available from respondents provides an important gain. It may not be possible to achieve a high level of explanatory rigour in positivist terms, but this can be regarded as a cost outweighed by the benefits of increased interpretative power. As we saw in Chapter 2, workers in some schools of qualitative research regard the concept of statistical rigour as quite irrelevant to their research work in any case.

Workers in our local field of interpretative geography see their discipline as being concerned with the interaction of the interviewer or observer and the observed as individuals within a mutually understood order, rather than the imposition of order on human behaviour by 'scientific' methods. Interpretation in this narrower sense can take many forms. In one corner we have ethnographic researchers who regard interpretation as the most appropriate method but still want to put their findings into a framework of scientific order, and in another are structuralists and semioticists much more concerned with symbols and images in the 'text' of the interview than the underlying realities that others might perceive (Eyles and Smith 1988: 2). If there is a unifying factor it is a concern with context. Ethnographic and interpretative techniques stress the importance of context, whether we are looking at the individual in society or the way in which landscapes acquire coding derived from the societies that live in them.

Interviews

Qualitative interviewing is an important technique. It might be used in association with structured interview or questionnaire techniques. Interviews with key actors of the kind we saw earlier are often carried out in conjunction with formal questionnaire programmes. However interviewing also has a significant place in qualitative geography in its own right.

We have already seen qualitative interviewing techniques in the context of key actors, and this review gave us some insight into the nature of interviews in this framework. Significant points were the small numbers of respondents involved, the importance of flexibility in handling the interview process, and the importance of interaction between interviewer and interviewee.

All interviews can be said in broad terms to be part of a process of interpretation. The key difference between questionnaires and formal interviews of the type we have looked at so far and the less formal interviews of qualitative geography is the level at which interpretation takes place. In the case of formal interviews interpretation is typically done by the researcher after the interview, but the framework in which it is carried out will have been erected long before the meeting takes place. In informal interviews, one of the most important tasks is to identify appropriate directions to take during the interview in the process of eliciting information from the interviewee. In formal interviews and questionnaire work any subtexts or hidden agendas that emerge from the answers have to be ignored, or at best mentally noted for a later revision of the questionnaire.[12] Only open-ended questions allow the interviewee some freedom of expression, and even here there is little room for interaction.

The interactive nature of qualitative interviewing means that interviewing is itself part of the learning process for the researcher. This means that we have to adopt a methodology that will let us take advantage of this process. Cook and Crang have pointed out that the 'read, then do, then write' approach typical of positivistic research is dangerously inflexible in this context. Ethnographic research will typically progress simultaneously on more than one front, with the researcher following up issues brought up in interviews and incorporating the results of this further reading in the next round of interviews. The research question itself might have to be recast (Cook and Crang 1995: 19).

Suppose that we are interviewing people about their attitude to the development of a new retail complex. We may have envisaged that the key issues will be matters like the range of choice, impoverishment of local facilities, and the logistics of organising shopping trips. As we build up our picture of attitudes and expectations from successive interviews, we discover that our first expectation was not quite appropriate. Shopping turns out to have cultural importance beyond simple issues of the mechanics of choice. Some shoppers, particularly the elderly, turn out to have strong feelings about less tangible issues such as images of place. Because we were able to give ourselves the flexibility to follow up these lines as they emerged, we will not have lost the information these respondents gave us. However, we will be prompted to go back to the literature and read more widely in fields such as cognition and landscape, and the psychology of the elderly before we return to the interview programme.

Informal interviews (sometimes called depth or in-depth interviews) are typically conversational in style. There will normally be a written or mental checklist of topics, but the nature and sequence of the questions will not be pre-defined. There is also an emphasis on recording the interviewee's own words, rather than simply allocating the response to a category. If it seems appropriate the interviewer can direct questions into areas that may not have been on the checklist. An alert interviewer can carry out what Oppenheim calls 'traffic management', diverting the interview into particular areas in response to indirect or direct prompts in the interviewee's answers. Open questions are typical, and in a fluent interview there may be very little need for questions at all (Oppenheim 1992: 72–73). You should not be surprised if some of the questions you ask turn out to be 'wrong', in the sense that they provoke unexpected unease or hostility. Although you should not make the mistake a second time, the first experience can be put to use. You can learn useful things about the interviewees' attitudes (and your own) by evaluating these responses.

Qualitative interviewing is not as concerned as questionnaire surveying with issues of number and representativeness. Instead of the sampling frames we looked at earlier, we are likely to focus on what is called 'theoretical sampling', in which our main criteria will be quality and positionality. Our primary concern is to interview people who have distinct and important perspectives on the theme of our research question. They might be chosen in advance. Some studies (particularly where members of the target group may be difficult to locate by normal means) might proceed by the 'snowball' method in which the first respondents located are asked to recruit their successors. The actual size of the interview group is a secondary issue. Rather than trying to achieve a target number of interviews, we will be alert for the stage when we have explored the whole range of realistic responses. Beyond this 'point of theoretical saturation', further interviews simply provide us with further examples of attitudes and behaviours we have already explored (Cook and Crang 1995: 11–12).

It should be emphasised that qualitative interview work is not an easy option. The mechanics of making contact and organising the times and places for interviews will be time-consuming, and the interviews themselves are likely to be much longer than formal interviews or questionnaire sessions. You will find a useful guide in Cook and Crang (1995: 37–40). Quick fixes will not work. Any attempt to conduct long informal interviews by intercepting shoppers on draughty street corners will be doomed to failure! The researcher has to be able to establish a good rapport with the respondents and it takes a lot of skill and experience to conduct an informal interview well and get useful results from it. Although a relaxed and conversational atmosphere is important, the interviewer always has to be aware of the underlying purpose, and must ensure that answers to similar questions are elicited within the same framework of meaning. The issue of 'intersubjectivity', the interaction of the different subjective frameworks of the interviewer and interviewee, always has to be taken into account.

Student researchers find qualitative work intuitively attractive but difficult to implement. One of the major problems from the student's perspective is that there are few of the defined procedures typical of the quantitative approach. Not only is it much more difficult to learn to use techniques effectively in the absence of clear-cut rules, but managing the study process effectively in qualitative fields like this requires a lot of experience and the ability to learn from it. One of the obvious steps forward for interested readers who have not yet got this experience is appropriate reading (e.g. Oppenheim 1992, Silverman 1993). Although most source texts in this field come from the social sciences, the selected readings in the texts like Eyles and Smith (1988) and Anderson and Gale (1992) are helpful in showing how qualitative geographers have made use of these techniques.

Even those whose preferences and research topics lead them to take a generally quantitative approach should acknowledge that it is important to be aware of the different forms that interviews can take. Depth interviews have a legitimate place in the piloting of questionnaires, and it can be argued that researchers who are aware of the range of responses that informal interviews can produce are in a much better position to write good questionnaires than those who have never explored territory outside the quantitative field.

Focus groups

A focus group provides a very different kind of resource, by assembling a number of people for a discussion concentrating on a particular topic – hence the word 'focus'. Whereas the interview explores individual attitudes in isolation, the discussion that takes place in focus groups will allow the researcher to see how people articulate, defend, and perhaps modify their views in interaction. If the group works well, this interaction will need little prompting, and it may provide insights that could not be guaranteed to emerge from individual interviews.

Managing a focus group obviously requires some care. Who should be the members? The membership of the group will have a vital effect on the way that its members interact, and putting together a group of participants needs thought. It might be easiest to use a group that already meets for its own purposes, such as a neighbourhood forum or pensioners' group. Apart from the advantage of being able to use established meeting arrangements, members of the group will already know each other and there will be no need to break the ice. However there are also disadvantages. In a group that meets regularly, discussion might be constrained by established pecking orders and entrenched viewpoints. It is also possible that community of interest will provide too much agreement on the issue in question for the discussion to be valuable. Assembling a group specifically for the purpose requires more work in contacting individuals, finding a suitable meeting time, and arranging a venue. However it does allow membership to be defined in a way that should promote discussion.

Members should as far as possible be chosen to be equal participants in the discussion. There should not be one member who can take on the status of 'expert' on the topic.

How big should the group be? If it is too small, it might not be able to provide a full discussion. However if it is too large the atmosphere may be intimidating, and the discussion can end up being dominated by one or two self-confident individuals. Generally speaking, groups with between six and twelve members probably work best, although the performance of the group on the day will depend on the topic and the individuals. The length of the session is equally flexible. Something between one and two hours is probably a good ballpark figure, although it is quite possible that discussion will dry up earlier or carry on well beyond the expected time (Cook and Crang 1995: 59).

What is the topic for discussion to be? It should obviously be relevant to the research theme, but it is equally important that it should be interesting to the participants. It need not be consciously controversial, but it must be one which people feel they own and which they think is worth discussing. How should the discussion be managed? Normally it will be controlled by a 'moderator', who will work from a checklist or set of themes, prompting discussion and acting as far as possible as a non-directive referee. The researcher may take this role, but it is probably better to ask someone else to take it on. This leaves you free to take an observer role and concentrate on the recording of the discussion. Neither researcher or moderator should be drawn into the discussion. They should take on the role of experts on procedure rather than on the topic itself.

Recording a focus group and analysing the task may be a major task. Only tape or video recording will suffice to record a lively discussion in which not just the verbal discussion but body language and visual prompts might be significant. The equipment should be good enough to capture all contributions and identify their source, without being so intrusive that it dominates the room and the discussion.

Because of the work involved and the level of input required from everyone concerned, focus groups are not to be undertaken lightly. Beginning researchers should get experience of interviewing before taking them on. However they have tremendous potential. While an interview programme should give us a good picture of individual responses to a development like a shopping centre, the interaction involved in a focus group discussion about the implications of the scheme will provide us with a view of the ways in which these individual views relate to each other in a social context.

Participant observation

Participant observation is the traditional method of field anthropology, in which the researcher gets access to a community, spends time living within it, then returns to the academic area to write the report. It has always posed the greatest problems of intersubjectivity. The term itself, with the implicit conflict between the roles of participant and observer, should make this clear. However it offers a great deal. Interviews and focus group discussions are artificial situations, and there is always the risk that this will affect the responses. As a fly on the wall, a participant observer has a better chance of recording attitudes and behaviour unfiltered by awareness of the outside world. However this particular fly cannot stay on the wall forever. To have any value, participant observation needs integration into the host community. It is important to get to know individuals well enough to understand how they interact with each other and identify key social actors in particular situations. It is quite normal for the subjects to know that research is going on, and they might be able to co-operate by keeping diaries, recording movements, and so on which will supplement the researcher's own field diary and notes.

Workplaces in particular are potentially attractive locations, since they allow a chance of observing communities interacting in one of their most important arenas, and not least because (as Cook and Crang have pointed out) they offer poor students a chance of paid employment while doing research (Cook and Crang 1995: 15). On the other hand employers are likely to be unsympathetic to research aims and intolerant of short-term workers who spend time visibly noting what is going on around them. This might mean that the researcher has to join the social circles of fellow employees, as well as doing all the recording out of working hours. Leisure activities in general provide a less precarious base for research than workplaces, although leisure is an intermittent activity and research might have to be carried out over a longer period than in the concentrated conditions of the workplace.

Not all research fields will offer the same opportunities. In the case of our retail example, taking a short-term job at a checkout might be the basis for good research on working conditions, but contact with a stream of customers in conditions that offer little chance of communication is not going to be helpful if we are studying customer attitudes. If we take the alternative strategy of attaching ourselves to a particular shopper community, we have to accept that size and fluidity of the customer base makes sustained contact difficult. We can focus on particular groups in a variety of ways. For example the major shopping centres like

Meadowhall in Sheffield attract a large number of coach trips from surprising distances. Travelling with the shoppers on these coaches might provide some insights, but only on a fairly superficial level.

It is fair to say that because of the high level of commitment needed by participant observation, it is less likely than other techniques we have already seen to fit into a composite research strategy. It is also less likely to be a suitable technique with which to start research. It poses a number of ethical questions more starkly than other techniques. As a researcher you might have to cope with information about anti-social or criminal behaviour given in confidence by respondents who are not entirely clear about your role. Your community might put you yourself in ethical difficulties. Cook and Crang suggest that as a general rule you as researcher should do what the community requires as long as it does not compromise your own values and beliefs (Cook and Crang 1995: 24). Following this rule may not be easy. A geographer of the author's acquaintance doing research on street gangs some years ago was put through a kind of initiation rite which involved sharing in a few criminal episodes. Not to do so would have put a major barrier between him and his community.

Other qualitative techniques

Film and video techniques have become popular in recent years as ways of studying social relationships. Existing photographs and videotapes can be used in their own right. They are not only images in themselves, but mute witnesses to the processes of selection by which the taker decided what was worth recording. As a result of this a collection of photographs can be used to interpret an individual's idea of what is significant. Most people take photographs with the intention of displaying them to their family and friends, so we can also reasonably expect that the images of people and places selected for recording will accord with the expectations of the photographer's community. Less commonly we may have access to films or videos made by special interest groups. Here the significance is likely to be both explicit and conscious. You will find a very useful discussion of this in Rose's analysis of two films made by local groups in East London (Rose 1994).

A more active way of exploiting filmic evidence is to get individuals to make photographic or video recordings of themselves, their activities, or their daily routines, as part of the research project. These can then be analysed. The popularity of projects such as the *Video Diaries* project run by the BBC means that a great many people are now familiar with the idea, even if they have not had the motivation or equipment necessary to produce their own. From the student's point of view, the disadvantage of this approach is related to the cost of hardware. Even providing research participants with cheap disposable cameras would make a formidable hole in a small budget. Video recorders are still expensive, and it would be a very trusting university that provided a student with video equipment to be handed out on loan to members of the public.

Less commonly groups and communities have been involved in mapping exercises. Common Ground, an environment and arts group, launched what it called the Parish Maps Project in 1987. A small number of maps created by recognised artists became the basis of a touring exhibition designed to engage community interest but the more important long-term effect of the project was the creation of about 1,500 community maps. Their content and appearance was very varied, not least because the processes of social negotiation and contestation during their production took very different forms. Some of them made explicitly political statements about local identity and planning issues. Others have to be read more carefully to elicit their meanings (Crouch and Matless 1996). Techniques like this can be adapted to a wide range of local uses.

So far we have looked at ways in which we can get information directly from people. It is also possible to work indirectly, by analysing artefacts, the objects created by human activity. Generally speaking geographers working in this field have concentrated on two areas of interest, texts and landscapes.

Textual analysis has a long history. The field of exegesis – the interpretation of religious scriptures – is an ancient one. In the recent past we have seen the development of quantitative techniques such as content analysis, which has been used throughout the post-war period as a way of standardising the analysis of recurrent patterns in texts. However, in our context a more important development has been the growth of a new interest in the construction and deconstruction of texts as artefacts with meanings beyond their explicit aim of communication. Analysis of a wide range of documents and public statements will show us that they are not transparent communication devices but have their own distinct cultural loading. Silverman (1993: 13) provides a useful introduction to this. Analysis need not be confined to formal documents but can be extended to a range of 'texts' like paintings, maps, advertisements, or posters. As part of our retail example we could analyse the key words and constructs used in documents for public consumption such as press releases, news stories, and council minutes. The loading of these texts might in itself be used as the basis for discussion in the focus groups discussed earlier. As a complementary study we could analyse the content of filmic records of the kinds discussed above.

Increasingly geographers treat landscapes as phenomena which demonstrate cultural processes in action. Useful introductions to research in this field include Massey and Jess (1995) and Anderson and Gale (1992). Treating landscapes as text is a uniquely geographical technique, and one that has been used a great deal in recent years. Since the landscapes in which we live are essentially human artefacts, they offer a large number of messages about their creators. Research work in this field has looked at landscapes both as artefacts in their own right and as the structured arenas in which human behaviour takes place. Because work in this area is so diverse it is best here to point the reader in the direction of a number of examples worth looking at as guides to the way in which this can be done. An example of the study of landscape as cultural artefact is Monk's (1992) work on the ways in which the built landscape can be interpreted in terms of gender roles, while Cosgrove *et al.* (1996) have examined at the cultural significance of landscape in relation to major reservoir schemes. Smith's (1993) study of seasonal fairs in the Scottish Borders examines the role of urban landscapes as venues for the expression of local social tensions at one level, and solidarity against the outside world at another. In a similar way May's (1996) work on Stoke Newington used interview techniques to explore the senses of place owned by different parts of the community and the ways in which class-related identities intersect and conflict. There is a sizeable body of literature on the ways in which landscape and culture intersect. Useful examples include Daniels and Rycroft (1993) on Nottingham in the novels of Alan Sillitoe, and Cresswell's (1993) study of Kerouac's *On the Road* as an examination of mobility in American popular culture.

Summary

The intention of this chapter was to demonstrate the range of techniques available for getting data from people, using the theme of a research project on retail development to emphasise the essentially complementary role of different techniques. On balance the emphasis has tended to be more on questionnaires and structured interviews, but this should not be seen as a reflection of their relative importance. One simple reason is that in geography these techniques are often introduced to beginning students as a fairly easy entry into the field of data generation. Another is that their use within a quantitative positivistic framework means that they require a high level of initial design and are bound by a set of well-defined conventions which require explanation. Returning to our retail example, it should be clear that many of the qualitative techniques available are less rule-bound, and much more depends on the acquisition of experience. Although you may meet them relatively early, the level of commitment and time needed for techniques like participant observation and the use of focus groups mean that you may not apply them seriously until you start to carry out individual research in earnest.

One final point must be made here. All research involving people as participants has an ethical dimension. Research in these fields is potentially intrusive and must be done responsibly. Normally it can only proceed with the assent of the participants, of course, but there may be a question about the nature of the agreement. In principle at least participants should know exactly what the aims of the exercise are, but this might not always be appropriate. In the retail case, for example, interviews with company representatives might only be forthcoming if the research aims are presented in such a way that the gatekeepers' suspicions are disarmed. However care has to be taken over this. If the questions turn out to indicate an agenda different from the stated one, the participant has every right to withdraw and refuse to let any information be used. A researcher should respect this right, and has little choice in any case if the person is a key actor who would be easily identified from the final report.

Confidentiality is important. A general rule is that the report should if possible reveal no information that could be linked directly to individuals, and certainly not without their consent. Nor should a report criticise identifiable people or make them seem foolish. What you see as raw data may be sensitive personal information to the respondent, and should be treated with respect. As we saw earlier, this can be a particular problem with participant observation, in which the researcher is likely to learn a lot from individuals in contexts where they may have relaxed their guard. What happens if people divulge information about anti-social or criminal behaviour? Should there be a point beyond which the researcher has a duty to society rather than the research project? If so, should people be warned when this point is being approached, even if this creates barriers? A decision will have to be made, and a lot will depend on the researcher's own ethical code as well as the context.

Discussion questions and tasks

1 Consider the problems of obtaining a sample of cyclists for use in a survey of attitudes to motorists as road users.

2 You have made contact with a neighbourhood group in an area affected by new retail developments. What techniques might you use with this group in exploring the impact of development?

3 There are at least three things wrong with this question. What are they? 'Do you utilise public transport regularly, and what do you consider its primary disadvantages?'

4 Here are statements given by two different respondents to an interpretative interviewer. 'In my early twenties I stopped using buses and bought a car.' 'In my early twenties I bought a car and stopped using buses.' What differences of emphasis are evident here and how might you direct further questions in the two cases?

Further reading

See also

Research and project design, Chapter 2
Data, Chapter 3
Statistics, Chapters 5 and 6

General further reading

Here is a short list of texts that provide useful and accessible information about the methods discussed in this chapter, and illustrations of the contexts in which they are used.

Oppenheim, A.N. (1992) *Questionnaire Design, Interviewing, and Attitude Measurement*, London: Pinter.

Silverman, D. (1993) *Interpreting Qualitative Data*, London: Sage.

de Vaus, D.A. (1996) *Surveys in Social Research*, London: UCL Press, 4th edition.

Cook, I. and Crang, M. (1995) *Doing Ethnographies* (CATMOG 58) Norwich: University of East Anglia.

Eyles, J. and Smith, D.M. (eds) (1988) *Qualitative Methods in Human Geography*, Cambridge: Polity Press.

Anderson, K. and Gale, F (eds) (1992) *Inventing Places: Studies in Cultural Geography* Melbourne: Longman Cheshire.

Notes

1 It may be important to ensure that the respondents do not know whether or not they are part of the control group. The classic case is the double-blind procedure used in drugs testing to ensure that neither the patients nor medical staff know which patients belong to the control group being given the neutral 'placebo' treatment. If the medical staff know which is which, they may tend to vary the treatment in response or behave in such a way that the patient becomes aware which group he belongs to. In our shopper case, the danger is that the more sophisticated of our non-car-owners might deliberately over-represent the effect of the transport campaign as a means of lobbying for its extension.

2 In common usage the term 'population' includes everyone living in a territory, irrespective of age or gender. In this context a population will usually be narrowed down to include the total number of individuals who have one or more attributes in common, although they will always be part of a population in the more general sense. The total number of women of working age in Somerset is a population in the narrower sense. So is the total number of people who watched the Eurovision Song Contest in May 1996. A population can also consist of occurrences. In statistical parlance it is quite normal to talk about populations of air journeys or shopping purchases.

3 There are exceptions. It is possible but not common that a researcher might persuade an organisation to oblige members of its staff to co-operate with a research project.

4 With the exception of the special case of longitudinal surveys, which we will look at later.

5 The formula is:

$$n = \left(\frac{Z\sqrt{PQ}}{C} \right)^2$$

where n is sample size, Z is a constant representing the chosen confidence level, P is the percentage for which the confidence level is computed and Q is 1–P. C is the desired size of the sampling error, called 'confidence interval' in the source text. A 'worst-case' scenario is commonly used in which $\sqrt{PQ}$ is replaced by 0.5. The formula assumes a very large population. This method is described in more detail in Haring *et al.* (1992: 65–67). Brave hearts can try Moser and Kalton (1971: 146–152). Tables of sample sizes for different sample errors are available. See for example de Vaus (1996: 70–73).

6 I am grateful to Peter Dalby of the Hertfordshire Beekeepers Association for the beekeeping information. Readers should not assume from this example that the beekeeping community is eager to be sampled by student surveyors!

7 Normally the results of the pilot survey will be discarded when their significance has been assessed. The surveyor will have taken the precaution of ensuring that the surveyed group for the pilot will not be part of the defined sample for the full-scale survey.

8 We can usually make an exception here for interviews with 'experts' such as planning officers or the organisers of pressure groups, but even here we should not expect too much.

9 This is not a sexist assumption. Most of the respondents in shopping surveys *are* female!

10 Some American store chains are said to have used private detectives many years ago to track their rivals' customers to their homes. This may or may not be true.

[11] This author has fond memories of a student undertaking a dissertation on the distribution of migrants from the Caribbean in inner London, who encountered so many closed doors that in the end he determined which islands his sample came from by identifying the dialects in which his unseen non-respondents told him to go away.

[12] In a properly piloted survey these should have been detected and allowed for at an earlier stage.

5 Coming to terms with statistics

Statistics demand respect but should not invoke terror. This chapter tries to put them in a proper context and covers:

- Probability
- Samples and populations
- Descriptive and inferential statistics
- Significance
- Probability and the normal distribution
- Non-parametric statistics

Statistical techniques are tools that all geographers should be able to use. However, over the years they have acquired an ideological significance that gives them quite a different status. For some academic geographers the use or avoidance of statistics has come to be a statement of methodological purity. This leaves students confused. Most dissertation supervisors have had meetings with distressed students who felt that their work would not be given full marks unless they made use of statistics, but could not find a meaningful way of doing so. Most dissertation supervisors can probably also think of examples where a student's choice of topic and research design seems to have been guided by a wish to avoid statistical testing! As always, the issue is one of appropriateness. The first decision must always be whether or not the use of statistics is necessary. Only if it is does the choice of technique become an issue.

Why should we use statistics? For most geographers there are probably two reasons. The first is that they allow us to attach numerical values to observable relationships between data. Statistical testing provides a way of assigning a level of confidence to our assessment of a relationship. The second reason is that more advanced statistics (largely beyond the scope of this chapter) allow us to detect and assess relationships which may not be directly observable. At this level the use of statistics becomes more than simply a means of confirming judgements made non-numerically.

Probability

Statistics is a science of probabilities rather than certainties, and its language reflects this carefully weighted assessment of outcomes. In statistics we accept or reject hypotheses rather than speak confidently about truth and falsity. Many statistical techniques are concerned with the relationship between individual events or cases and general patterns of behaviour. If a researcher asks one person a series of questions about leisure activities, the results may be interesting but can hardly be used as a basis for generalisation. However if the same researcher then goes on to ask the same questions of another 199 people, general patterns will certainly emerge. Another look at the original case at this stage might show that the individual's responses turn out to be typical of the population in some ways, but quite distinctive in others. Statistics recognises that the members of a population are similar but not identical. A population can be described both by measures of central tendency which indicate the characteristics most typical of it, and measures of dispersion, which show the range that individual members might occupy.

The theory of probability is concerned with the likelihood that events will have particular outcomes. Certain results can be eliminated entirely. Current knowledge says that it is impossible for anyone reading this book to live for ever, so the probability of that outcome can be set at 0 per cent or 0.0. In the same way, since every person must die, the probability of any individual's eventual death is 100 per cent or 1.0. But when will it happen? The probability that someone will live to be a centenarian is a small but real one. Insurance companies, for which issues like this are significant, maintain actuarial tables based on real observations which allow them to calculate the probability of this taking place with some confidence for members of a particular population. However to calculate at birth the probability that any person will die at a particular age would be a futile exercise. Far too many factors affect the outcome of the individual case.

One of the classic demonstrations of the workings of probability is coin tossing. Tossing a coin has two possible outcomes, heads and tails, and these have an equal chance of occurring which can be set at 0.5 or 50 per cent. Every time a coin is tossed the odds of it coming up tails are 0.5, because individual tosses are independent events and do not affect their successors. Before writing this section the author tossed a coin twenty-five times with the following (quite genuine) results:

TTTTHHHHHHHTTTTTHTHTHHTTHHHH

The overall balance of heads and tails was 13:12, as close to equality as the odd number permits. However there were four runs of at least four heads or tails, and little sign of simple alternation. This illustrates quite neatly the basic nature of probability. The order in which individual cases occur may be unpredictable. but as the number of occurrences increases it is likely that predicted outcome and reality will converge.

When outcomes are independent, the odds of an outcome are not affected by any previous cases. If you win the major prize in the National Lottery, your chance of winning it with the same numbers the following week is unchanged.[1] In other cases events are dependent on others and this affects probability. There are sixty-four clubs in the draw for the third round of the English FA Cup. Before the draw starts the probability of one particular club being drawn against any other is 1/63, or 0.0159. The sum of these probabilities is 1.0. However if that club has still not been picked when all but four of the balls have been selected, the probability of meeting any of the other three unpicked clubs is much higher at 0.333. At this stage the probability of playing any of the other sixty clubs in that round is of course 0.0, so the total probability still sums to 1.0.

In cases where the range of outcomes is clearly defined, it is possible to work out the probability of sequences occurring. We have seen that there is a probability of 0.5 that a coin will fall showing tails. What is the probability of a specified sequence, like TH or HH? Since this is a combination of independent outcomes, we calculate it by multiplying the individual probabilities. For either of these outcomes the probability will be 0.5 × 0.5, or 0.25. This can be confirmed easily enough, since there are only four possible outcomes of tossing a coin twice – HH, HT, TT, and TH. Any one of them has a quarter of the total probability, or 0.25. For longer runs the probabilities are obviously smaller. The likelihood of matching the run of six heads shown in the example above is only 0.016, less than 2 per cent.

In the case of real-world data the position is often less tidy. Suppose that we have decided to test the reasonable hypothesis that attendance at football league matches directly reflects the success of the home team. Table 5:1 summarises data collected to test this hypothesis, namely the frequency

Table 5:1 Attendances at Endsleigh Insurance League Division Two matches 1994/95

Attendance	Frequency	P
0 – 2,999	69	0.125
3,000 – 5,999	337	0.611
6,000 – 8,999	88	0.159
9,000 – 11,999	29	0.053
12,000 – 14,999	10	0.018
15,000 – 17,999	9	0.016
18,000 – 20,999	8	0.014
21,000 – 23,999	0	0.000
24,000 – 26,999	2	0.004

with which attendances at Endsleigh Insurance League Second Division games fell into different classes in the season 1994/95. A total of 552 games were played, and of these 337 fell into the 3,000–5,999 category, giving an overall probability of 337/552, or 0.611. The other probabilities have been calculated on the same basis, and as we would expect their sum is 1.0

However we need to know more about these probabilities. Less than half of 1 per cent of all games had crowds of more than 24,000. Is this true every season, and is it purely chance that no games at all had attendances in the next largest category? Looking at the pattern as a whole, can we take this group of figures as a set in its own right, or are they only members of a bigger population that might behave in a different way? Is there a regularity in the way that this pattern is distributed that might let us make inferences from it? We will pursue questions like this later in this chapter.

Samples and populations

All statistical tests are carried out in relation to populations of data. A population is simply a body of data to which an individual case can legitimately belong. However the majority of tests are performed on samples rather than whole populations. The example in Table 5:2 might help make the difference clear. Table 5:2 shows small fictitious sets of data describing the spending characteristics of three different groups of people using shops in an out-of-town shopping centre.

Since a superstore will have many thousands of customers during a typical week, it should be obvious that none of these columns can claim to describe the total population. Because of this they must be treated as samples. It is not common for a research worker to be in a position to gather data about a whole population. In a case like this one the stores will continue to add to the populations of shoppers as long as they stay in business. Much the same would be true if the data collection involved

Table 5:2 *Sample weekly superstore spending patterns of three groups (£)*

A	B	C
51	55	51
47	52	26
45	54	40
53	59	74
37	47	5
68	63	66
58	60	66
61	61	52
39	46	6
49	56	14
47	55	48
53	57	40
48	51	29
50	58	30
43	50	53

traffic surveys, or spatial behaviour on tourist beaches. For this reason, and because all data gathering exercises have to operate within limits of time and budget, samples are the norm. The most important question that must always be asked about a sample is whether it is representative. If it can be accepted as such, its characteristics can be taken to be true of the population as a whole. Otherwise it must be discarded.

In cases like Table 5:2 where we possess more than one sample we have to face a second question, whether or not they represent the same population. So far in this case we know nothing about the way in which they were gathered. They might differ only in the date on which they were collected. On the other hand the different samples might have been designed to test the responses of distinct groups of people on the basis of characteristics like gender, income range, mode of transport to the shop, or the time of week when the trip took place. One of the most important sets of tests in statistics allows us to compare samples and establish whether or not they can be treated as parts of the same population. If we were to find that two or more of the samples above did come from the same population, the significance of this would depend on the context. This might have been the expected result. On the other hand if the samples had been differentiated on a variable like gender or income, membership of the same population would suggest that this variable was not a major factor in determining the size of the shopping bill.

As we saw when sampling techniques were discussed in the context of project design in Chapter 2, one of the factors that should be considered at the design stage is the range of statistical tests that might be carried out on the data. The form of random or systematic sample used in the data generation process will be vital in relation to the statistical analysis carried out at this level.

Descriptive and inferential statistics

Statistical techniques can crudely be divided into two groups – descriptive and inferential statistics. Descriptive statistics focus entirely on the sample gathered in the data generation exercise.

The measures of central tendency described in Chapter 3 (mode, median, and mean) are the most basic descriptive statistics. We can also add basic measures of dispersion in the form of maximum, minimum, and range. The values for our data sets are summarised in Table 5:3.

There is not a lot that can be said about these. Sample B shows the best correspondence between the central measures and has the narrowest

Table 5:3 *Descriptive statistics for the superstore samples*

	A	B	C
Mean	49.9	54.9	40
Median	49.0	55.0	40
Mode	47	55	40 and 66
Maximum	68	63	74
Minimum	37	46	5
Range	31	17	69

range. In that sense it is the most orderly data set, whereas Sample A has less consistency in central tendency and more dispersal. As for Sample C, the fact that it is bimodal (two numbers share the modal position) and has a much larger range than the others suggests that it lacks the internal consistency of the other two. However, descriptive statistics give us no reliable way of knowing how these samples relate to each other, or whether they might belong to the same larger population. They obviously offer us a limited horizon, and a large-scale survey handled only with descriptive techniques would be a waste of time.

Inferential statistics provide much more analytic power. They allow the researcher to use the sample to draw conclusions about the population to which it belongs, and make inferences about the characteristics that might be expected in other samples as yet to be selected from that same population. One of the important uses of inferential statistics is to take the converse approach, and test samples to see whether or not they can be regarded as members of the same population. This would be an obvious way of comparing the three shopper samples in Table 5:2.

However all power carries responsibility, and the greater power of inferential statistics must be used carefully. The user should always pay attention to the three guidelines below.

- Sampling must be independent. This means that the data generation method should give every observation in the population an equal chance of selection, and the choice of any one case should not affect the selection or value of any other case.
- The statistical test chosen should be fit for its purpose and appropriate for the type of data selected.
- The user must interpret the results of the exercise properly. The numerical outcome of a statistical test is the result of an exercise carried out in a rather artificial environment governed by particular rules. This has to be put in the context of the real world.

Without attention to these points, the results produced by tests like the ones described below might be not only wrong but dangerously misleading.

Significance

One of the most important concepts in statistics is that of significance. Significance is in essence the degree to which we can accept that an inference is supportable. We have already seen the use of the null hypothesis H_o. If we were comparing two of the samples in Table 5:2 to establish whether or not they came from populations with different means, the null hypothesis would be that there was no significant difference between them. If we were able to reject H_o this would mean that we would be justified in treating our samples as members of different populations. When the test has been performed and the results evaluated (for how this is done see Chapter 6) the outcome of the test will allow us to define a significance level. Significance is normally stated in terms of the fractional likelihood of the rejection being false, and the most commonly used levels are 0.05, 0.01, and 0.001. The level chosen will depend on the context. For many geographical applications 0.05 is quite acceptable, but in other fields like medical research a more rigorous level like 0.001 may be more usual.

No matter how convincing it is, we can never be completely confident that the characteristics of a sample are a perfect reflection of the total population, but significance levels at least allow us to define a degree of confidence. Suppose that we are able to reject the null hypothesis at the 0.05 level. This means that the difference between the samples is big enough to have occurred by chance only in 5 per cent of all cases if both samples had come from the same population. The other levels of significance are progressively more strict. At the 0.001 level there is only one chance in a thousand that the different samples could have come randomly from the same population. Most statistical tables simply allow us to determine which cut-off point has been reached, but statistical computer packages may give precise significance levels. A value might be displayed as 0.023 for example, showing that H_o can be rejected at the 0.05 but not the 0.01 level

This means that the testing of significance is an exercise in the application of probabilities, and does not give us absolute certainty. It is normal to select a target significance level before running the test, and this requires judgement. If the level is set too generously, there is a danger that we will produce a Type I error (see Chapter 3) by rejecting the null hypothesis when it should be upheld. However it is not always a good idea to go for the strictest possible level of significance, since this might lead to the risk of a Type II error, and the acceptance of a null hypothesis that could quite legitimately have been rejected. Reducing the risk of one type of error always increases the risk of the other. The general rule is that if there is any doubt, we should try to eliminate Type I errors. It is better to deny ourselves a genuine explanation than accept one that is false. In practice, working in a particular field will show the levels of significance that workers have come to accept as suitable within particular contexts. There is thus a body of precedence and 'case law' that allows the new researcher to make decisions.

Another important general principle is to make a clear decision about an acceptable significance level before the test is run, and adhere to it. It is not good practice to search for some significance level at which the null hypothesis can be rejected if the initial level does not allow rejection.

One of the issues that all newcomers to statistics are puzzled by is sample size, a topic we considered in a broader context in Chapter 4. How big does a sample have to be for effectiveness? Generally speaking, the larger the sample the better. Most statistical tests function more powerfully with larger samples. However the cost of collecting a large sample might be high, and the sort of small-scale low-budget work typified by a student research project or dissertation cannot generate large samples. From the student researcher's point of view, the ideal is to work with samples just large enough to yield acceptable levels of significance. However there is no simple guideline, and we will come back to this issue when we look at the different types of test.

Probability and the normal distribution

Earlier we saw that the behaviour of a population could be described in terms of measures of central tendency and dispersion. An important development from this is the ability to test a sample or population's behaviour against a set of expectations. Statisticians have spent a lot of time developing different models of expected distributions, and one of the most important of these is the normal distribution.

The normal distribution is based on the assumption that although it is not possible to predict exactly where any individual independent event contributing to a population will fall, the population behaves in such a way that its members will be distributed in a predictable and symmetrical way around a mid-point, which will be defined by the mean. A distribution curve can be fitted to this distribution, and will describe the outer boundary beyond which it is predicted that values will not occur. A normally distributed population is shown in Figure 5:1.

One of the most important elements of normality is the standard deviation, a measure of dispersion usually identified by the Greek letter

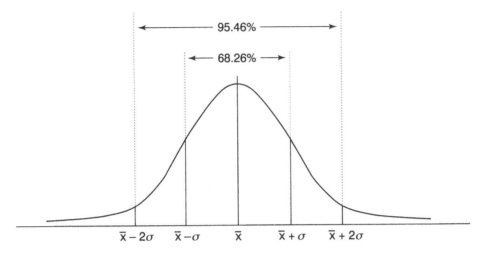

Figure 5:1 *A normal distribution curve*

sigma – σ. The standard deviation is calculated in a fairly straightforward way. The individual differences between occurrences and the mean are all squared, and the mean of the squares is calculated. This mean square variation is known as the variance. The standard deviation itself is the square root of the variance. The importance of the standard deviation is that for a normally distributed population it defines the proportion of cases that will fall within given distances of the mean. For a sample with the mean $\overline{X}$(note the bar over the letter), the distribution is shown below:

68.26 per cent of all occurrences fall within $\overline{X} \pm 1\sigma$
95.46 per cent of all occurrences fall within $\overline{X} \pm 2\sigma$
99.70 per cent of all occurrences fall within $\overline{X} \pm 3\sigma$

If we know the mean and standard deviation of any real-world example, it is simple to work out the values around the mean at which the boundaries will fall. The larger the sample and the better its match to the normal distribution curve, the better should the position of the boundaries match the projections above.

The significance of this is far-reaching. The relationship between mean and standard deviation gives every population a distinctive 'fingerprint'. Statisticians have developed a battery of powerful tests based on the normal distribution. Power in this context means the probability in using a test that the null hypothesis will be rejected appropriately, i.e. when it is false. These tests based on the normal distribution are sometimes described as parametric statistics, on the basis that the characteristics of a sample – sample mean, sample standard deviation and so on – are used to infer their true equivalents in the population as a whole. As we have already seen, we will not usually know the real values for the whole population; these population characteristics which can be defined even if we cannot assign fixed numerical values to them are parameters. Among other tests, the normal distribution provides powerful ways of testing whether single samples belong to particular populations, or whether groups of samples are similar enough to be assigned to the same population. There are tests based on the comparison of means, and others which evaluate dispersion by analysing differences in variance or standard deviation. We will look at some of these tests later.

Strictly speaking, parametric statistics should only be used where the samples can be regarded as belonging to a population governed by the assumptions of the normal distribution. How often is this likely to be the case?

Many samples do not fit the normal distribution as well as we might like. We can take as an example the football attendance case we looked at earlier. Superficially we might expect crowd sizes to behave in a way that approximates to normality. If we assume reasonably enough that attendance will reflect a team's success, we should expect the distribution to centre around the mid-table position, with both extremes defined by appropriate levels of enthusiasm.

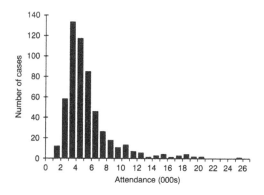

Figure 5:2 *Attendances at Endsleigh Insurance League Division Two matches 1994/95*

However if we plot the distribution of classes as a histogram (Figure 5:2), it is immediately obvious that this sample does not match the normal distribution well. There is a heavy concentration at the low end of the range, with a long tail of small numbers of occurrences to the right. The distribution has what is called positive skewness, with the majority of values being lower than the mean, which in this case is 5,512.

The distribution is also strongly leptokurtic. Kurtosis is the measurement of the degree to which values are concentrated in one part of the distribution, and a leptokurtic distribution is markedly peaked, with more concentration in one area than the normal distribution would support. The opposites of these conditions, by the way, are negative skew and platykurtosis, the condition in which the distribution is generally flatter than one would expect for the normal distribution.

If we are confronted by this type of pattern, a number of questions pose themselves. Is it acceptable to use parametric tests in a case like this? What would be the effects of using parametric tests inappropriately? What alternatives are there? Underlying these queries are other important questions. Why are these attendance patterns behaving like this? Might it be that our first simple assumption failed to identify the dynamics of the situation correctly?

The first question has a simple answer at first glance. Strictly speaking, parametric tests should only be used with observations drawn from normally distributed populations (Siegel and Castellan 1988: 20–21). To follow this advice rigidly would mean abandoning parametric statistics in all but very special cases, because in most instances we only know about a smallish sample, not the total population. Practical experience shows that it is very rare for samples to be really good matches to the normal distribution's expectations in terms of symmetry and skewness, but we have no way of knowing whether the population as a whole might conform better. In practice researchers tend to allow themselves to use parametric tests if sample behaviour does not deviate very much from the normal. Unfortunately, defining where the boundary might lie requires judgement and experience, and these cannot be acquired immediately.

Luckily we do not need a lot of experience to eliminate obvious problem cases. If we return to the football crowd example, this is certainly an example where we should not try to apply parametric tests. The histogram we looked at earlier indicated discouraging levels of skewness and kurtosis, and descriptive statistics confirm the problem. Our example has values of +2.54 for skewness and +8.24 for kurtosis, whereas the expectation for a normally distributed population would be 0 for both. The mean is 5,512 and the standard deviation is 3,357, and in the sample data sixteen cases (2.9 per cent) lie more than three standard deviations from the mean, against the expected 0.03 per cent. Since it is a large sample of over 500, it is probably not a bad indicator of patterns in the population as a whole, which confirms that it would be wise to look for other ways of testing other samples of this type of data.

The second question about the effects of using parametric tests inappropriately has no clear-cut answer. A test might work and produce results that look acceptable, but its power cannot be guaranteed. Statisticians have worked out how powerful different tests are when they are used within set rules. The problem is that there is no consistent way of telling how the level of power deteriorates when the rules are broken.

To apply a parametric test when there is more than a slight deviation from normality is therefore to step into the unknown. Even worse, statisticians have not been able to say confidently what a 'slight' deviation might be (Siegel and Castellan 1988: 21)! Faced with this difficulty, the student's best survival guide is to read existing work in the field and see which tests researchers have felt confident to use with the type of data being examined.

The third question is an important one. If we cannot rely on parametric tests in a particular case, what alternatives do we have? One point to note is that some populations conform to other distributional rules. Consider serious accidents (railway accidents, coach crashes, major fires, and so on) which put extra strain on medical services. If the occurrence of these is recorded on a weekly basis for a hospital catchment area, most weeks will probably have a zero score, and a smaller number will have a single case. We can expect that for every increment in the weekly score there will be progressively fewer cases, so that the pattern has a strong positive skew. Cases like this are sometimes fitted to a Poisson distribution, which assigns probabilities by applying an exponential constant to known behaviour. In our fictitious case the medical authorities can use the information as a basis for calculating a level of service to meet different levels of predicted risk. Unlike normally distributed data, populations or samples suitable for Poisson analysis are typically skewed, although it should be obvious that not all skewed patterns are suitable for Poisson analysis!

In other cases the best we can say of our sample data will be that they do not match either of these distributions. If our data are in ordinal or even nominal form, we cannot use the normal or Poisson distributions in any case, since these are only appropriate for interval or ratio data. In cases like these, our best course is to use one of the non-parametric tests discussed in the next section.

Before we move on, however, it might be worth thinking again about the football data. If we were to expect them to be normally distributed, it would be on the assumption that the number of supporters was strongly and positively related to the success of the home team. This was the basis of our tentative hypothesis. However there are two problems. The first is that we can expect some die-hard fans to defy logic and come to watch their team no matter how poor they are. The team which performed worst throughout the season 1994/95 generated the ten poorest attendances and a mean below 2,400, but attendance only once dropped below 1,500. The other issue is that visits by successful teams will usually bring large numbers of away supporters. Even a poor team will have a good gate when a well-supported team turns up, and the largest crowd of all (almost 26,000) was for a vital end-of-season match between the two leading teams. Taken together, these tendencies compress the range below the mean but produce a small number of very high values.

Looked at in this perspective, football attendance data are unlikely to be normally distributed. This simply reinforces the point made earlier, that in any research exercise, no matter how small, the nature of the data should be taken into account from the earliest stages. Anyone gathering data in a field like this should be aware of the problem, and ready from an early stage to use tests that will better fit the data available. The non-parametric tests introduced below will often be appropriate.

Non-parametric statistics

The term 'non-parametric' is used to cover a very wide range of tests. Their one significant common characteristic is that they are 'distribution-free'. In other words no assumptions are made about the distribution of values, and they are designed to work in contexts where the nature of the data would exclude use of parametric statistics.

What are the advantages of using non-parametric methods? Most of them provide results in the form of exact probabilities, which means that there is no need to worry about the shape of the underlying distribution. If the data are in nominal or ordinal form, their advantage is quite simply that they are the only legitimate tests. Tests on ordinal data are among the most useful in the range. Some non-parametric tests can be applied with very small sample sizes which are not admissible with their parametric equivalents, and some at least can be learned and applied very easily.

Non-parametric tests have sometimes been dismissed as 'soft', and it is true that the parametric tests tend to have higher levels of power-efficiency – the degree of explanatory power that can be obtained from a given sample size. This being so, if the data meet the requirements of the normal distribution model it makes more sense to use a parametric test, which will provide the same explanatory power from a smaller sample. However the non-parametric tests can be made to provide powerful results, even though it may be necessary to work with larger samples to do so. It is also always better to use a distribution-free test if there is any question about the appropriateness of the data for parametric tests.

Summary

Like all tools, statistics can produce powerful results but can be disastrous in unskilled hands. Before starting an analysis, be really clear about the hypothesis you are testing and the nature of the test. As we worked through the football attendance example, we discovered a series of problems. The data were not suitable for parametric analysis, and there were strong spatial ties which meant that location had to be taken into account. Most important, our initial hypothesis about home gates relating to the success of the home team turned out to be poor. Even a fairly cursory look at the structure of football crowds raised issues of relative success, the role of visiting supporters, and spatial linkage, which made it very unlikely that this hypothesis would be worth testing.

Discussion questions and tasks

1 The terms 'probability', 'normal', and 'significance' all have special meanings in statistics. Are you confident that you know what they are?

2 Returning to the football league example, in what other ways might the culture of football invalidate attempts to make simple statements about crowd sizes?

3 Use this experience to evaluate the problems of making statistical assumptions about a field of study of your own choice.

4 Investigate the non-parametric tests. What advantages do they have to compensate for their lower power?

Further reading

See also

Data, Chapter 3
Sampling and sample sizes, Chapter 4
Statistical tests, Chapter 6

General further reading

Ebdon, D. (1985) *Statistics in Geography*, Oxford: Blackwell, 2nd edition. This is a good basic text for geographers, written in a clear style with a lot of examples. At a more advanced level there are several texts which have stood the test of time. These include:

Siegel, S. and Castellan, N.J., Jr (1988) *Nonparametric Statistics for the Behavioral Sciences*, New York: McGraw-Hill.

Hammond, R. and McCullagh, P. (1978) *Quantitative Techniques in Geography: An Introduction*, Oxford: Clarendon, 2nd edition.

Blalock, H.M. (1979) *Social Statistics*, Tokyo: McGraw-Hill – Kogakusha, 2nd edition.

Notes

[1]   We are looking here at the wins as individual outcomes. The probability of winning the jackpot twice running is very much smaller!

6 Statistics in action

With an understanding of the framework within which statistical tests operate, we are now in a position to see how some of the more widely used tests can be applied. We will look at:

- The t test for independent samples
- The t test for paired samples
- The Mann-Whitney U test for independent samples
- The Wilcoxon signed-rank test for paired samples
- The Chi-square (χ^2) test for independent samples
- Tests of association
- Independence and autocorrelation
- Statistics and the computer

In the previous chapter we looked at the framework within which statistical tests are applied. In the sections that follow we will move on to look at a number of parametric and non-parametric tests designed to carry out a series of commonly required tasks. The aim is not to provide a description of the working method, which any basic statistics textbook will provide. The priority here in is to explain the underlying logic of the test and illustrate how its results might be interpreted. Consistent use of the same data set allows us to make a preliminary comparison of the relative strengths of the different tests.

In the first place we will look at comparisons between different sets of data. The null hypothesis will be that the two data sets compared are randomly selected samples from the same normally distributed population, or from two populations with identical behaviour. When we test to see if the null hypothesis can be rejected, we are thus establishing whether two samples can significantly be assigned to different populations.

To provide a working example we will return to the fictitious shopper data from the earlier part of the chapter, concentrating on the first two data sets. Although these are not perfectly normal in distribution, they can be taken as approximating to normality. We have already seen their basic descriptive statistics. In addition to that, both have acceptable kurtosis scores around 0.64, although sample A is less skewed at 0.286 than sample B at 0.601.

A   51 47 45 53 37 68 58 61 39 49 47 53 48 50 43

B   55 52 54 59 47 63 60 61 46 56 55 57 51 58 50

When we compare two populations, our experimental design might take one of two forms. Very often the samples are simply independent groups of observations, and the test will concentrate on the behaviour of the samples as wholes. There is no need for the samples to be identical in size, although they should not differ grossly. Sometimes the sampling procedure might be so designed that each observation has a counterpart in the other sample, in which case it would be appropriate to treat the observations as being paired. To make our fictitious example fit the bill, we will assume that the observations fall in pairs occurring in the same order. For example the 68 in Sample A and the 63 in Sample B will be a pair. Pairing will normally be done if the two populations contain members between which there is some functional link. A common situation which we will simulate later is one in which the paired observations are 'before' and 'after' responses from the same individual. For obvious reasons paired tests need identical population sizes in the two samples.

For the purpose of the tests of difference we are going to look at, we will assume that a superstore has launched a campaign to attract shoppers without access to cars. These will tend to be people who tend to shop little and often to avoid carrying too much weight. The campaign has involved a well-publicised improvement to the courtesy bus service and assistance in loading shopping onto the bus. In this way the management intends to make it easier for these shoppers to buy and transport more goods per trip, thus improving profits and strengthening customer loyalty. In both cases our researcher has surveyed bus-using shoppers before and after the improvement. In the case of the paired samples we will assume that the same shoppers have been interviewed before and after the event.

The t test for independent samples

One of the most useful parametric tests in this field is the t test (sometimes called Student's t), which can be applied in independent or paired form. Since the null hypothesis in this case is that there is no significant difference between the two samples, we therefore expect the values to be distributed around the same mean. Broadly speaking, the t test divides the difference between the means (measures of central tendency) by a measure of dispersion. The closer the difference between the means is to zero, the more difficult it will be to reject the null hypothesis. For independent samples the test statistic t can be calculated by this formula:

$$t = \frac{\overline{X}_1 - \overline{X}_2}{\sqrt{S_1^2/N_1 + S_2^2/N_2}}$$

$\overline{X}_1 - \overline{X}_2$ is the difference between the two means. The two N values are the sizes of the two samples, and S_n^2 for each sample is the variance.

When the formula is worked out the value of t emerges as 2.04. This means very little until it is assessed against a significance table, which simply lays out significance assessments for different values of t. An important element here is the issue of 'degrees of freedom'. The degrees of freedom (DF) value is a function designed to ensure that the test takes account of the number of occurrences involved. In cases where sample sizes are identical it is $N - 1$, but if the sample sizes differ (or might do so) it has to be calculated by a rather complicated equation. In this case the DF value works out as 23.36, which can be rounded down to 23. In most significance tables each DF value will be presented alongside a series of values of t relating to different levels of significance. In our example with DF of 23 a t value of 1.71 would be sufficient to reject H_0 at the 0.1 level but t would have to be 2.07 for us to reject H_0 at 0.05. We are just unable to reject the null hypothesis at this level. Using a computer program we can get a significance level of 0.053 for our example, confirming how close to rejection we have come.

The t test for paired samples

As mentioned earlier, it is sometimes possible to tie observations in two samples together in the form of matched pairs. If the samples behave in a genuinely distinct way there should be a consistent difference between pair members. The formula below describes a method calculating the value of the statistic t for paired data.

$$t = \frac{\overline{D}}{S_D / \sqrt{N}}$$

In this case $\overline{D}$ is the observed difference between the means, S_D is the standard deviation of the differences between members of pairs, and N is the sample size. The formula looks simpler, but conceals a good deal of work and is best done using computer techniques.

For the purpose of this test we will change our assumption, and suppose that our fictional researcher has interviewed the same respondents twice to obtain 'before' and 'after' data for individuals. The results of applying the paired formula are quite different. The DF value is N – 1, or 14, and the calculated value of t is 4.98. This allows us to reject H_0 at every level down to and including 0.001. In this case we can be confident that we are dealing with different populations. As far as the research topic is concerned, there appears to have been a positive response to the advertising campaign.[1]

We must be careful about the way we evaluate this result. It is derived from data which simulate genuine pairs, and the use of the second version of the test allows us to concentrate on this relationship. However it would be foolish to apply the paired version in every case in the hope that it would yield a more convincing result! If there is not a paired relationship, the results may in fact be worse than under the standard method. To demonstrate this the paired test was run on a modified version of the same file. The actual data values remained the same, but the values in one of the two samples were randomly shuffled to break the pair relationships. This time the significance level was 0.083, meaning that H_0 could not be rejected even at the 0.05 level. This result was marginally less significant than with the unpaired test.

The Mann-Whitney U test for independent samples

Many of the non-parametric tests are not only able to handle data that do not accord with the normal distribution, but are designed to work with ordinal or even nominal data. One of the more useful in the comparison of independent samples is the Mann-Whitney U test. This test requires only that the observations should be random, and that they can be ordered. The Mann-Whitney test will often be used with ratio or interval data (this is necessary because an absolute scale is needed to relate the rank series of the two samples) but these will be converted into a ranked series for the purposes of the test. If we want to apply it to our retail data we will have to recode them in ordinal form.

The Mann-Whitney test is based on the principle that if two samples are part of the same population, then when the samples are pooled and ranked the members of the two samples should be evenly distributed through the rank series. If the two samples show clustering within the ranked series they may not belong to the same population, but if the rank series of the samples behave in a similar way H_0 will have to be upheld. Since we have to recode our interval shopper data into ordinal form, the simplest way is to write out our thirty values in ascending order with an indication of the sample to which they belong. The ranks can then be defined (special rules govern tied cases) and the test carried out. The fundamental principle is to work through the sequence, summing the number of times that the value for one sample precedes a value for the other. The summed ranks are used to calculate the Mann-Whitney U statistic. The way in which the statistic is actually calculated differs according to the number of occurrences, and you should consult a specialist text for advice. There will always be two possible values, one for each sample group, and the smaller value of U should be chosen.

When the test is run on the shopper data, the mean ranks of the two samples turn out to be 12.5 and 18.5 respectively, and the value of U is 68.0. Inspection of the appropriate significance table shows that we are not able to reject H_0 at the 0.05 level. The actual probability value is 0.0675. A correction is available which takes account of tied ranks, and if we use this correction it is possible to reject H_0 at this level. Some computer packages will perform the correction automatically.

The Mann-Whitney test is more universal than the t test, since it is not bound by the assumption of normality. Although there is little detectable difference between the outcome of the two tests for independent samples as applied to our shopper data, the Mann-Whitney test's explanatory power is generally rather smaller, since the conversion of actual values to ranks reduces the amount of information available, but it does not present the same risk that comes from using the t test on data that violate its assumptions. The guiding principle is that if there is any doubt about normality, Mann-Whitney is the safer test to use, but it might be useful to take larger samples to compensate for its lower power-efficiency.

The Wilcoxon signed-rank test for paired samples

There are also non-parametric tests appropriate for paired samples, and one of these is the Wilcoxon signed-rank test. This test is not particularly powerful, but is easy to use and relatively fast. Like Mann-Whitney, it is suited to use on interval data, and converts it to ordinal form. It works on the principle that if the two samples belong to the same population, there should be no significant difference between the behaviour of the pairs. However, if the actual values show a consistent direction with (say) the first members of the pairs tending to have higher values, the test should be able to detect a significant difference.

To apply it to the fictitious shopping data, we once again make the assumption that our two samples provide us with fifteen matched pairs.

The test is simple to carry out. The difference between the scores in each pair is worked out, and the differences are ranked (tied cases are given the average of their ranking positions). When the ranking has been done, each rank is given a positive or negative sign according to the direction of the difference. The signed ranks are separately summed and the smaller sum used to give us the test statistic T, which can then be checked in a significance table. In our case almost all the differences are positive. There is one tie and only one negative case, which gives us a T score equivalent to its rank of 5.5. This allows us comfortably to reject H_o at the 0.01 level.

The Chi-square (χ^2) test for independent samples

The last test of comparison we will look at in this context is the Chi-square or χ^2 test. This is one of the most widely but not always wisely used of the non-parametric techniques, and can be applied to single samples or for comparison. It has two features which make it rather different from the other tests surveyed so far. First, it uses category or nominal data. Second, the test can be used to compare actual data with a pattern of expectations set up by the user.

To demonstrate it we will change the rules for our fictitious data set again. In this case we will assume that sample A has been taken among a group of non-car-users, and sample B among car-users. Because Chi-square needs category data, the spending values are converted from interval to category mode simply by dividing them into those below £50 and those of £50 or above. All the 'observed' shoppers can therefore be allocated to one of four groups forming cells as shown in Table 6:1. Our 'expected' distribution is based on a null hypothesis that car ownership has no effect on shop spending. At first glance this would seem to indicate that all four cells should have the same contents, i.e. 7.5 cases each. However this is not actually the case. Two-thirds of all the observed cases spent at least £50, and we must build this into our expectations. We can calculate the expected value for each cell simply by multiplying the row and column

Table 6:1 *Observed and expected frequencies*

	Observed		Expected	
Car-user	2	13	5	10
Non-car-user	8	7	5	10
	< £50	> = £50	< £50	> = £50

intersect totals and dividing them by the overall total. For example, the value of 5 in the top left cell of the Expected set (car-user and £50) is derived by multiplying the column total (10) by the row total (15) and dividing it by the global total of 30.

The Chi-square statistic itself takes a number of forms, of which the one below is among the most familiar. For each cell the difference between O (observed) and E (expected) is squared and divided by the E value for the cell. The sum of all the cell operations gives us χ^2. The value in this case turns out to be 5.4. Degrees of freedom in this test will be the number of rows minus one (1) times the number of columns minus one (1). DF is therefore 1. In this case we can reject H_0 only at the 0.5 level.

$$\chi^2 = \sum \frac{(O - E)^2}{E}$$

Chi-square is not difficult to use even with small tables like the 2 × 2 example. However, like all tests it has to be used carefully. One problem is that it is unreliable if the expected value in any cell is very small. It is unwise to set up a table in which expected values of 5 or less occur in more than 20 per cent of cells, and in a 2 × 2 table like the example there should be none. Nor should there be any cells with expected values of zero. In our example it would have been more informative to be able to divide the spending values into three or more categories, but this was not possible without at least one cell having an expected value below 5. Another point to note is that with 2 × 2 tables some statisticians think that a modification called Yates's correction for continuity should be used. Applied to our example, it would reduce the Chi-square value to 3.75 and would prevent us rejecting H_0 even at the 0.05 level.

Tests of association

So far we have looked at tests which have one common objective. They all test the likelihood that samples belong to the same population. However there are also useful tests which test association. That is to say, they look at the degree to which we can see how the behaviour of different data sets is related. The simplest tests of association are correlations, which test the degree to which variations in pairs of observations mirror each other. Correlations produce coefficients in the range between −1 and +1. A value of +1 is a perfect positive correlation, meaning that there is a perfect linear relationship between the two sets of observations. Every increase in a value in one population will be matched by an exactly proportionate increase in the corresponding value in the other population. A value of –1 means a perfect negative relationship, with every increase matched by a proportionate decrease in the other member of the pair.

To provide an example of this we can return to our football example and add another dimension. We already have information about attendances. We will now add population figures for the host towns. Table 6:2 includes raw data from the 1991 census, modified where necessary by simple assumptions. In cases where large cities support more than one club (Birmingham, Bristol) the total population is divided by the number of clubs. In the London case the rather unrealistic assumption is that the population of each club's host London Borough forms the catchment.

Table 6:2 *Endsleigh Insurance League Division Two: average gates 1994/95 and population 1991*

Club	Gate	Pop'n	Club	Gate	Pop'n
AFC Bournemouth	4,391	154,677	Hull C	4,721	253,111
Birmingham C	16,983	312,588	Leyton O	3,436	203,343
Blackpool	4,771	145,175	Oxford U	6,148	124,058
Bradford C	6,152	450,708	Peterborough U	5,055	149,402
Brentford	6,536	196,602	Plymouth A	5,832	244,163
Brighton and H A	7,563	143,356	Rotherham U	3,278	247,776
Bristol R	5,173	186,044	Shrewsbury T	4,013	91,318
Cambridge U	3,443	101,643	Stockport C	4,525	278,645
Cardiff C	4,543	277,182	Swansea C	3,582	182,577
Chester C	2,388	115,680	Wrexham	4,071	113,862
Crewe A	4,239	102,231	Wycombe W	5,856	154,907
Huddersfield T	11,665	369,172	York C	3,685	101,436

One of the most commonly used correlations is Pearson's product–moment correlation. The Pearson test works by calculating the product of all the deviations from the mean for the pairs of values in samples X and Y then dividing it by the product of the two samples' standard deviations (S_x and S_y) multiplied by a population function. In terms of our example, X and Y represent the gate and population figures for each club. The formula looks like this (it can be written in other ways) and yields the Pearson statistic r.

$$r = \frac{\sum (X_1 - \bar{X})(Y_1 - \bar{Y})}{(N-1)S_x S_y}$$

From the description it should be clear that this is a parametric test, and we have already seen that our football attendance data do not fit the assumptions of normality, so we must treat our results with suspicion in this case. When we have carried out our calculation the value of r is +0.4934. There is a positive correlation between football attendance and town size, but not a very impressive one. Like the other tests we have looked at, we can assess significance. We can reject H_o at a one-tailed significance level of 0.01 in this particular case.[2] In other words, we can be confident that there is only a relatively small probability that this value of r was generated by chance. Be aware that we are saying here only that the correlation coefficient is likely to reflect reality, and not that it is actually strong!

The most common non-parametric correlation is Spearman's Rank correlation coefficient. As the name suggests, it is based on an assessment of the difference between the ranks of occurrences in the two samples being compared. In Table 6:2 Cardiff City, for example, are thirteenth in crowd size but fifth in population, giving a difference value of 8. The formula looks like this, where d is the difference in ranks, n is the sample size, and r_s is the Spearman correlation value.

$$r_s = 1 - \frac{6 \sum d^2}{n^3 - n}$$

In this case the coefficient value for our example is +0.4400, indicating a correlation which is still positive but rather weaker than in the Pearson case. Checking a significance table shows that we can reject H_o at the 0.025 level in this case. This might give us only a little more confidence in our result at first glance, but if we remember how poorly these data fit the normal distribution assumptions we will be aware that this non-parametric test is more appropriate, and more likely to give a reliable result.

In this particular context we have to acknowledge that even if population size helps explain the size of football attendances, it does not do so very well. Perhaps we have simply measured population size in a clumsy way that does not reflect the real distribution of potential visitors. Perhaps others factors we have not looked at must be taken into account. We must always keep in mind the purpose of our use of statistics, and the environment in which our results have to be seen.

Independence and autocorrelation

All the tests we have looked at, parametric and non-parametric, require that observations should be independent. However we must not assume that the populations from which we draw our samples behave in ways which guarantee this independence. If we come back to our first football example, that of crowd sizes, we have an example where we must question the independence of the occurrences. The major problem relates to near neighbours.

We should not expect crowd size simply to reflect the success of the club itself. Geographical factors have to be taken into account. An ailing team with a successful neighbour may lose support faster than one with no local rival, since some supporters who prefer to watch a winning side will transfer their allegiance to a successful team if one is accessible. Local 'derby' matches between neighbours often attract large crowds. All but five of the teams in Table 6:2 had their highest home gate against teams which were local rivals, championship contenders, or both. The derby element on its own accounted for about one-third of top gates. In addition the geographical distribution of fixtures is managed by the fixture organisers in association with the police authorities, in order to prevent crowd trouble when matches take place simultaneously at neighbouring grounds.

For these reasons we might question the independence of the observed results. We are dealing here with an aspect of something called spatial autocorrelation. Values are spatially autocorrelated if they cannot be treated as a random distribution over a geographical surface. Observations taken close to each other may show a definite association. Indeed most geographical research would be pointless if spatial patterns of this kind did not exist. If we were sampling crime rates in a city, for example, we would be very surprised if our sample showed random distribution rather than a pattern with distinct local variations. There is a growing body of spatial statistics designed to measure and explain the way in which phenomena behave geographically.[3]

Returning to our football example, our observations are not autocorrelated in the direct way that crime rates might be, but there can be no doubt that there is a strong spatial element in their occurrence. Unless we recognise this, our statistical analysis is likely to be faulty and even misleading.

Statistics and the computer

The calculation of statistics on paper can be slow and error-prone. Fortunately there are ways around this. Many pocket calculators are programmed to calculate standard deviations and other descriptive statistics. Computers offer a wider range of facilities. Spreadsheets allow their users to perform descriptive statistics, and dedicated statistical packages like SPSS and Minitab provide a range of inferential statistics which includes everything this chapter has described, and much more. If you become a serious users of statistics you will almost certainly want to explore these facilities. Not only do they save time and reduce the chance of error in routine calculations, but they offer facilities which hand calculation could not realistically provide.

Although computers have revolutionised the handling of statistics, it is sensible at this point to make it clear that there are some problems with computerised statistics. The first is that the use of the computer brings 'overheads' with it. One of these is the time cost of familiarisation. The training needed to make a statistical package do exactly what you want may take far more time than the hand calculation of a simple test! Even a seasoned user might find that in the case of simple tests it is faster to use a pocket calculator and paper than to set up the data and command files needed by the package.

A less immediately obvious problem is that of 'correctness'. A statistical computer package will normally provide an error message if the user has made a mistake in presenting the data or writing the list of commands. The naive user might be forgiven for assuming that when the package produces values for test statistics or significance levels and generates no error messages, these results must be right. Unfortunately, this is not necessarily true. There are many ways in which a careless or inexperienced user can make a package produce unwanted or defective results without breaking the rules and thus producing error messages. More serious than that, current packages have no critical faculties. If the user asks the machine to use an inappropriate test, the results will normally be produced without comment.

The computer should never be treated as a black box which magically produces correct results. As a user, you should be confident that the chosen test will be appropriate before any computer analysis is carried out. When the analysis has been performed, it is important to check the working to make sure that it matches expectations. If the same analysis is to be performed on several sets of data, it is good policy to carry out a hand calculation on one set to ensure that the instructions given to the package have been appropriate.

Summary

Performing a statistical test is only part of the research process. As was said earlier in this chapter, any test you choose should be fit for its purpose and appropriate for the type of data selected. However, even if the test is well-chosen, the ultimate success of the exercise depends on the choice of an appropriate hypothesis before the statistical stage, and an appropriate analysis after it.

Discussion questions and tasks

1   The front end. Define a simple statistical task involving a small data set and select a test that seems to meet your needs.

2   The back end. When you have done this, evaluate your results. Can you understand the logic underlying the test, and do the results look correct from a 'common-sense' perspective?

3   Integrate. Are you clear about what you have done, and was it the right thing?

Further reading

See also

Data, Chapter 3
Sampling and sample sizes, Chapter 4
Concepts and general issues in statistics, Chapter 5

General further reading

Ebdon, D. (1985) *Statistics in Geography*, Oxford: Blackwell, 2nd edition.

This is probably the best text to use when you are learning about statistical tests. To take your work to a more advanced level you should use one of the texts already listed in Chapter 5. These are:

Siegel, S. and Castellan, N.J., Jr (1988) *Nonparametric Statistics for the Behavioral Sciences*, New York: McGraw-Hill.

Hammond, R. and McCullagh, P. (1978) *Quantitative Techniques in Geography: An Introduction*, Oxford: Clarendon, 2nd edition.

Blalock, H.M. (1979) *Social Statistics*, Tokyo: McGraw-Hill – Kogakusha, 2nd edition.

Notes

[1]   If we assume that no other factors have intervened! In a properly designed before-and-after paired study we would have used a control group to allow us to be confident that the change did not stem from some other more general factor.

[2]   We have ignored the difference between one-tailed and two-tailed tests so far. Two-tailed tests allow for the probabilities at both ends of the normal distribution. We have used two-tailed tests so far, because we could not be sure in which direction one sample might differ from another. In this case, knowing that we have a positive correlation, we are justified in looking at only one end of the distribution.

[3]   A useful introduction to this field is Chapter 7 of Ebdon (1985).

7 Maps and mapping techniques

Geography is not just about maps, but they are still vital to us. This chapter explores:

- The purpose of mapping
- Fundamental concepts – scale, direction, projection, and co-ordinates
- Do maps tell the truth?
- Maps for purposes
- Professional map production
- The student as cartographer

Maps are a familiar part of our environment, but most users are not conscious of the forms that maps take and the rules that bind them.

The purpose of mapping

Maps have traditionally been one of the geographer's primary tools and most geographers have training in map-making, although the production of high-quality finished maps is normally done by specialist cartographers. All maps are attempts to represent patterns in geographical space, but as we will see later there is a very wide range in the types of patterns portrayed and the methods used.

Broadly speaking all maps conform to three rules. The first is that they are dimensionally systematic. The ways that spatial relationships on any map are handled will be based on some set of principles, although the result may not always have a very obvious relationship to visible ground reality. The second is that maps have to represent three-dimensional patterns on flat surfaces. The ways in which this is done will vary. Finally, all maps are selective. The map can show only a limited set of data from the vast range that might be applicable to the area mapped.

While it might be tempting to think that these three guiding principles are sufficient, we have to take into account three more important issues before we can go much further. The first is that maps do not represent absolute and total truth about whichever part of the earth's surface they portray. All maps are the creations of cartographers who are deliberately selective in the information they display. The second point is that maps have to be interpreted. Cartographers use a special language of symbolisation to communicate information, and a user who does not know this language will not get a full or correct message. Finally, the way in which the cartographer designs the map and the way in which an individual user assesses it are governed by matters of human perception. The brain's response to visual images is still not fully understood.

All in all, then, maps do not simply reproduce information. They pass on selective messages, and they do so in a coded form that must be decoded for successful use, using a medium that we still do not fully understand. Even the most mundane of maps contains a lot of room for misinterpretation. The training and experience of professional cartographers enables them to work comfortably within a set of familiar conventions about how information is communicated. It tends to be forgotten that map users do not have this training and experience. Using a map is not intuitive. For example cartographers in the past regularly depicted buildings and other objects with oblique-view pictograms. The modern convention is to use a vertical plan view throughout, and some users find this level of abstraction difficult. Significantly enough, the older convention survives now mainly in maps designed to be 'friendly' and attractive to non-specialists. Graphicacy – the ability to read graphic images constructively – is a learned skill like numeracy and literacy, and uneducated map users may have a very poor awareness of what the map is meant to show and what the patterns actually represent.

Maps are designed to work within specific contexts. They are tools designed for particular purposes, and the terminology used in the field reflects this. Navigational maps are often described as charts. The term 'plan' is used to describe detailed maps precise enough to mean that measurements taken from them can reliably be used in physical planning or civil engineering contexts. Highly stylised maps departing from conventional representation are often described as diagrams. It is also traditional to recognise a distinction between topographic ('general' in American usage) and thematic mapping. Topographic maps provide a general-purpose picture of the earth's surface, mainly but not exclusively depicting visible surface features. Figure 7:1 uses a fictional landscape to show a typical topographic map panel using the conventions of the British Ordnance Survey. Thematic cartography produces maps designed to show distributions and relationships within particular areas of study. Thematic maps may show highly abstracted patterns like the geography of educational scores or nationalistic attitudes.

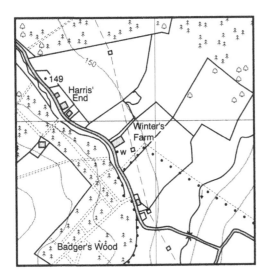

Figure 7:1 *A typical topographic map panel*

Cartography has shared in the intellectual debate that has affected all academic disciplines in the post-war period, although it would probably be fair to say that the discipline has accepted ideas from outside rather than being at the leading edge of intellectual development. This is not the place for a detailed interpretation of the way that this development has taken place, but readers who want to be better informed about it will find a useful discussion in the work of MacEachran (1995: 1–16). However a short summary will help provide a perspective.

Until quite recently cartography was a rather uneasy combination of scientific or mechanical processes operating within a positivist framework, such as the construction of projections, and the 'artistic' skills which were meant to produce an attractive and readable final product. When cartographers like Arthur Robinson began to explore issues relating to perception in the 1950s, they believed that it would be possible to extend the positivist framework. They thought that they would be able to establish laws of visual logic allowing cartographers to be confident about the users' response to particular choices of symbol, colour, or shading. Although the artistic elements of design would not be eliminated, better understanding of the rules of perception would provide a more rational basis for design.

In the 1960s and 1970s, attention turned to graphic communication, and the emphasis was on information transmission. In communications models the assumed aim of the cartographer (or at any rate the person on whose behalf he was working) was to transmit a particular message through the map. Maps were seen as information channels, and a lot of attention was paid to eliminating the distortion of the message or 'noise' resulting from weak map design.

This approach has been criticised on a number of grounds. One is that it assumes a strongly directed message within the map. However, a large number of maps are designed simply to present information. Although the information presented on a population distribution map, for example, will be affected by value judgements and issues of academic interpretation, the map itself is intended to provide an information base, and the user will analyse it rather than receive a single message. This is even more the case if we look at the ordinary topographic map. Topographic maps may depict contentious patterns – disputed political boundaries, for example – but their overall purpose is descriptive rather than propagandist.

In the recent past much more attention has been focused not on the map itself, but on the 'pre-map' stage of decision-making and choice, and the 'post-map' stage in which the user extracts and evaluates information. This does not mean that the importance of the map itself has been devalued, but simply that it is recognised as one stage in a complex process of data generation, presentation, and analysis. MacEachran describes his approach, which is not necessarily one shared by all working cartographers but does show the current direction of thought, as cognitive-semiotic. The cognitive element recognises the continued importance of studying the science of visual perception and visual inference. The semiotic side opens up a much broader context, by examining the role of 'meaning' in cartography. At its simplest, it is the study of 'how maps represent via signs' (MacEachran 1995: 214). This is an approach based more on the communication of meaning than the communication of messages.

Fundamental concepts – scale, direction, projection, and co-ordinates

To carry out their task of representing the real world, maps have to make use of a series of fundamental spatial concepts. At the simplest, these are distance, direction, and relative position. These are employed quite differently on different maps, and usually in formally stated ways.

Distance is most often represented through the device of scale. Scale is the mathematical relationship between the real world and the map. It can be thought of most simply as the ratio of a distance shown on a map to the real-world distance it represents. If the scale is 1:50,000, for example, each centimetre on the map represents 50,000 cm, or 0.5 kilometre. The larger this 'representative fraction', the smaller the scale of the map. A world map small enough in scale to occupy a single atlas page may be on a scale like 1:165,000,000, but detailed plans are routinely published on scales as large as 1:1,250, and individual maps on even larger scales are not uncommon.

Direction poses a problem on a spheroid like the earth, since any systematic reckoning of direction needs fixed reference points.

Fortunately the earth rotates, and this has provided the basis for directional systems based on the rotational and magnetic poles. The display of direction is often standardised by the use of orientation. Orientation means that the map is set out in such a way that the user can easily relate the directions shown to those of the real world. By long-standing convention, maps are laid out with north at the top of the image, although the term orientation itself harks back to the time when European maps were directed eastwards, towards Jerusalem. Two of the most familiar elements of map furniture are therefore a scale bar providing a visual display of reference distances, and a north point confirming orientation.

Consistent measurement of scale and direction opens up the possibility of building a reference framework which allows every point to be assigned a unique location. The science of the classical world was aware that the earth was a globe, and developed the framework for the modern system of earth surface measurement, based on a co-ordinate system consisting of 180 degrees of latitude from north to south, and 360 degrees of longitude on an east–west basis. Any location can be defined in terms of latitude and longitude. Important as this rather clumsy system is for global navigation, localised map work usually substitutes a Cartesian grid of arbitrary lines. The National Grid of the British Ordnance Survey, for example, is a hierarchical system which allows points to be defined to different levels of precision. The National Grid uses letter codes to define 100 km-square blocks. TL is such a block on the northern outskirts of London, and TL30 defines a particular 10 km-square block within it containing most of the Broxbourne area. TL3601 refers to a single square kilometre in Cheshunt and TL360012 locates the 100-metre square within which most of this chapter was written, using a keyboard which (if such a level of precision was necessary) is in the square metre defined by TL3604501264.

The combination of these working principles creates a rather reassuring model of clear relationships, in which there is a constant relationship between distances, areas, and orientations on the map and their real counterparts. Sadly, the simplicity of this model does not survive a comparison with reality. Cartographers face the problem that all earth surface maps are attempts to portray parts of the curved global surface on a flat plane, something that cannot be done without deformation. Although the flat earth model of the Christian cosmography temporarily eliminated the difficulty, it had to be confronted again when the idea of the spherical earth returned in the sixteenth century.

Over the centuries since then a great many map projections – at least 250 – have been produced to solve this problem. Until quite recently geography students were trained to be proficient with projections, and even now some awareness is necessary. All projections must be compromises, and most of them have tended to emphasise accuracy either of direction or scale. The so-called conformal projections are designed to provide consistent accuracy in directional terms. A navigator's course following an angular bearing will appear as a straight line when plotted on a conformal map. There are obvious benefits in this, but it is done at the cost of consistency in scale. An alternative is to use an equivalent or equal-area projection, which preserves distance or area relationships, but cannot be used to calculate direction. Some of these have to be displayed as 'interrupted' projections, in which segments of the earth's surface are shown like the peeled skin of an orange.

The relevance of this issue extends far beyond the technicalities of map production. The world views of individuals and societies have been influenced by the type of map available. One of the most familiar of all projections is Mercator's conformal projection, first published in 1569. In this projection area is progressively enlarged towards the poles, and this has given generations of users a quite incorrect idea of the size and shape of the earth's main land masses. The most visible case is Greenland, which widens spectacularly on Mercator maps towards the north. In fact deformation on Mercator maps is so great at high latitudes that they are usually cut off about 80° north, and northern Greenland never appears at all! More significantly, Mercator users are given a false idea of area relationships throughout the map. Africa, visibly much smaller on the Mercator map than North America, is actually about 25 per cent larger. Brazil and Alaska seem to be more or less equivalent in size, but Brazil is actually more than five times as large as Alaska.

The scientific cartography that developed on a large scale in post-medieval Europe incorporated Eurocentric mental frameworks, and the later rise of the United States reinforced the emphasis on the mid latitudes of the northern hemisphere. The persistence of the Mercator projection discussed above illustrates this rather well. Recently there has been more awareness of the need to escape from the constraints of this world view. Projections have been designed specifically to provide different perspectives. One example is the Peters projection, designed by Arno Peters in the late 1960s. This is an equal-area projection which divides the world into a grid of 10,000 rectangles and maps them in correct area proportions. The political significance of this was emphasised in 1981, when UNICEF distributed large numbers of copies of a publicity leaflet incorporating a world map based on this projection. UNICEF's choice of this projection was based on the desire to get away from a Eurocentric world view and to emphasise the size of the surface area

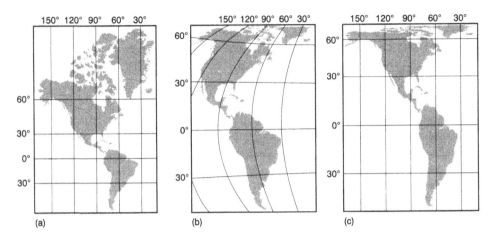

Figure 7:2 *Three projections of the same area*
(a) Mercator projection; (b) Hammer Equal Area Projection; (c) Peters projection

of the Third World. More recently a world atlas has been published using the Peters projection.[1] Figure 7:2 shows the striking difference between the images of the Americas as shown by Mercator, conventional Equal-Area, and Peters projections.

Although the Peters projection keeps area relationships, it deforms distances. In real life Africa is about 10 per cent longer from north to south than east to west. On the Peters map it is compressed to become more or less half as wide from west to east. Linear scale is therefore quite different depending on the direction of measurement. The Peters map is one example of an increasing use of cartography to challenge traditional views by choosing new projections, different focus of attention, and different orientation. However, we have to recognise that the nature of projections means that they are trading new deformations for old.

The majority of professionally produced maps are projection-based. For example, although it may not be immediately evident, all the products of the British Ordnance Survey are based on a locally defined Transverse Mercator projection (in which deformation increases west–east rather than north–south). The choice of this projection means that inaccuracies are small across the whole projected surface, and insignificant within the areas covered by individual maps.

On all maps which handle distance and direction consistently, relative positions will be defined automatically, allowing us to assess which areas are adjacent to each other, or the order in which points fall along a line. However, not all maps follow cartographic convention in the way they depict real-world scale and distance relationships. In such cases relative locations come to the fore. Primitive sketch maps may be wildly inconsistent in handling true distance and direction but satisfactory from the user's point of view if the user simply needs a representation of adjacency relationships or sequences of places. There are also cases, as we will see later, where advanced cartography chooses to shed its normal conventions and concentrate on relative position.

Do maps tell the truth?

The projection problem illustrates one of the difficulties of map production. Any projected map distorts the surface it represents. The distortion may be controlled and systematic, but it will still produce a specific set of deformations that will affect the user's perception. Any map depicts a version of the truth rather than an absolute truth.

The problem affects maps' content as well as their framework. All maps are highly selective documents.

This can be illustrated by comparison with the role of the photograph. In the pioneer days of aerial photography during and just after the First World War, the new medium looked likely to supersede the map. A mosaic of photographs could be used to produce a working plan of a potential battlefield within hours, a speed of response which traditional cartography could not match even when the territory was not in hostile hands. However, in the longer term photography came to provide support for the map, rather than direct competition. One of the major reasons for this is that the photograph is unselective.

Photographs simply record what is visible. They faithfully record surface textures and transient patterns which the cartographer may not want, but they have weaknesses in important fields. They cannot depict the abstract and invisible. A photograph of an urban area cannot record abstractions like political boundaries or the functions of buildings. It may record the surface textures of a park in great detail, but will not show the underground railway running beneath it. Only the top surfaces of multi-layer features will appear, but the cartographer is likely to want to show other patterns entirely – the roads and paths that run through woodland, rather than the treetops that conceal them. Photographs are almost useless in the field of thematic mapping.

In contrast to the photograph, all maps are selective. The needs of clarity and fitness for purpose mean that the cartographer must eliminate unwanted information. How this selection is carried out will vary a great deal, and will be determined mainly by the planned purpose of the map.

The commonest form of selection is simply to exclude categories of information that are judged to be irrelevant. However there is a subtler form of selection carried out by all map-makers, sometimes without conscious awareness of what they are doing. This is generalisation. According to Buttenfield and MacMaster, 'a typical objective of generalisation is to capture the essential character of some phenomena and remove unnecessary spatial and attribute detail' (1991: 124). At all scales, information has to be simplified for presentation. Even if space permits, the cartographer will not try to map every tile on a roof or every sleeper on a railway track. In the unlikely event that the original survey included this level of detail, the cartographer would probably suppress it as being distracting rather than informative. There are no hard and fast rules for the way this simplifying process is carried out. It may be done with little conscious thought by the cartographer on the basis of training and experience, the quality of the survey, and the purpose of the map.

As map scales change and space becomes more limited, the task becomes more challenging. Mapping a railway yard at a scale like 1:2,500, the cartographer has ample room to depict individual tracks and points, lineside buildings, and other features. At smaller scales the amount of space is obviously much reduced. At 1:2,500 even quite a small station yard will occupy several centimetres square of map space, but at 1:50,000 only a few millimetres will be available. To cope with the problem the cartographer has to come to terms with the related issues of symbolisation and scale-related generalisation. Figure 7:3 shows the amount and type of information shown in exactly the same fictionalised square of urban landscape on three progressively smaller scales. It shows very clearly how little information can be shown as scale decreases, and the large station symbol in Figure 7:3c also illustrates the use of symbolisation.

Symbolisation abandons the pretence of depicting visible truth, and substitutes standard symbols for representations of reality. Typically the symbol set will be displayed on a reference key accompanying the map. Some symbolisation is necessary at all scales, since almost all maps have to display abstract information without physical correlatives. Text labels can be used for this, but at smaller scales they are likely to become cluttered and illegible. Symbolisation comes into its own at this stage, packing abstract information into a compact space and providing visual impact. If we come back to the railway yards of our previous example, we can see how symbolisation changes with scale. At the 1:1,250 or 1:2,500 scales, the British Ordnance Survey maps railway layouts in detail. Simple box symbols first appear at 1:10,000, although there is still some attempt to show the detail of track layout. By 1:50,000 the majority

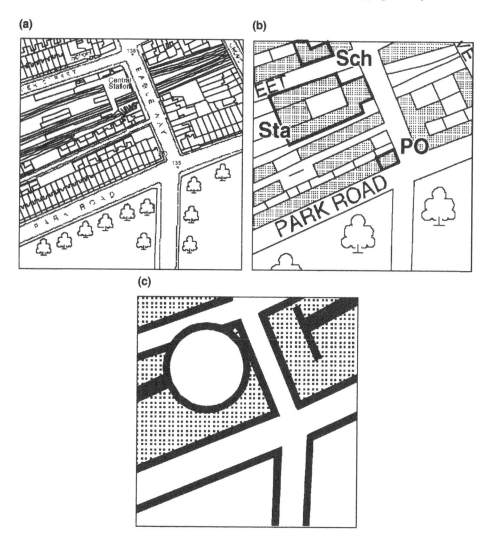

Figure 7:3 *The effect of generalisation*
(a) 1:1250 scale; (b) 1:10000 scale (c) 1:50000 scale

of stations are shown by red dot symbols, although there is an attempt to show the ground plans of the largest. At the really small scales like 1:250,000, even this distinction has gone, and all stations are shown by identical dots. On specialised maps symbols can be made to work much harder than this. Some transport maps, for example, use symbols structured to give a high level of information about station status, interchange facilities, and even opening hours.

Symbolisation is an ancient and intuitive feature of mapping. However the structured use of symbols is relatively recent, and the way in which they are perceived and understood is at the heart of the modern debate discussed earlier.

Although generalisation is always part of map production, it has to be confronted most obviously when it is forced on the cartographer by change of scale. Given the task of reproducing a map on a scale smaller than already exists, the cartographer has to recognise that reduction in the available area will mean that the information-carrying capacity of the map is reduced. The process of generalisation by which this reduction is carried out is one of the most difficult issues in map production and handling. As scale is reduced, the relationship between real-life and map distances becomes more strained. To give one trivial example, it is not difficult to calculate that the parallel lines used to depict motorways on the British Ordnance Survey's 1:250,000 scale maps indicate a road width of about 250 m, as against 40 m or so in real life. Narrow country roads on the same map are 100 m wide, as against 7.5 m or less in real life. The example helps illustrate two general principles. The first is that object or symbol size is exaggerated to provide clarity, and this will tend not to be consistent. The smaller the object, the greater the exaggeration necessary to display it adequately. The second is that as selected objects expand to consume more map space, there is less room for other information. There is inevitably a tension between the objectives of displaying information and maintaining clarity of image. Sound and consistent decisions have to be taken about what is included or excluded.

A number of different factors come to bear. All generalisation processes have to take into account the practical limitations of the graphics process. Whether the map is to be printed or displayed on a screen, we cannot shrink lines and dots beyond certain limits without creating ambiguity. Even if the technology available allows detail to be shown at very high resolution, the limitations of the average eye's ability to resolve detail might make it unwise to use the full level of resolution possible.

However, the most important constraints on generalisation relate to purpose. Most mapping agencies recognise that the user groups who use the different scales on which they publish will have different needs, and the maps must be designed to cater for these. Generalisation almost always involves a high degree of symbolisation, and it is inevitable that as scale becomes smaller and the amount of information shown per unit area drops, the level of symbolisation increases. Only the information most relevant to the map's perceived purpose can be included, and it will be symbolised rather than directly derived from ground truth. A mapping agency maintaining coverage on a series of different scales has to recognise that generalisation will radically change the appearance and content of maps at different scales.

This is a challenge. In fact generalisation has been identified as one of the major bottlenecks in moving map-making into the computer age. Even now the planning and detailed execution of the generalisation process is still done largely on the basis of experience, although progress is being made towards the design of expert systems that will take on the task (Buttenfield and MacMaster 1991). It is worth thinking a little about the nature of the challenge, since the decisions it requires are at the heart of map design. It is vital to map the entities or objects that the purpose of the map demands, and the attributes or qualities that are appropriate. However they cannot stand in isolation. Their spatial relationships must be preserved. This means that as far as possible relative distances and directions are maintained, and the map should be accurate in showing the connectedness of points and the contiguity of areas (Buttenfield and MacMaster 1991). Simply changing the scale of an existing map is not enough. Some agencies have experimented with photo-reduction to reproduce individual maps at different scales, but legibility suffers and information unwanted at the smaller scale is not edited out. Not surprisingly, this labour-saving method has not been widely adopted.

Cartographers have always recognised that the information shown on a map should show fitness for scale as well as fitness for purpose. The most effective map for a particular purpose on a particular scale will almost certainly have been designed with fitness for that scale and purpose in mind, and it might be surprisingly different from a map produced by the same agency from the same base survey, but with a different market and different scale.

So far we have considered deviation from truth as an inevitable part of the map-making process. It should also be said that throughout history maps have deliberately been used as vehicles for particular viewpoints, in extreme cases presenting blatant propaganda. The way in which this has happened is explored in a number of texts including Barber and Board (1993) and Wood (1993). Strictly speaking this has less to do with cartography in itself than with our next topic, the use of maps for particular purposes.

Maps for purposes

All maps have defined purposes, and these will determine the map's appearance and content. We can approach this issue by looking at maps in the transport field. Figure 7:4 shows three different types of transport-related map. Aeronautical charts are designed for international navigation, and must work within agreed conventions to avoid confusing

(a)

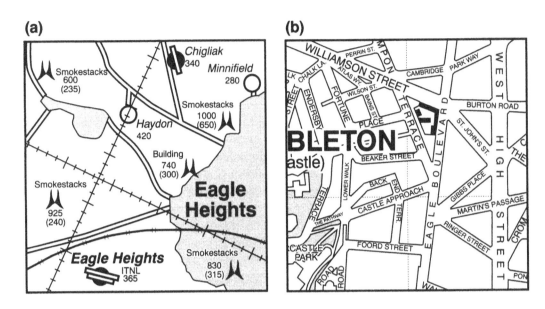

(b)

(c)

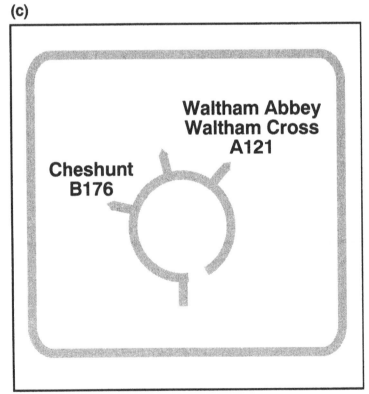

Figure 7:4   *Maps for different purposes*
(a) fictional aeronautical chart showing typical information such as obstructions, airfields, and identifying features such as major roads and railways; (b) fictional street atlas showing the priority given to street identities; (c) a typical road sign

their users. Their navigational role makes it important to use conformal projections, and because orientation to compass north is important for navigation they always use the traditional north-based orientation. However, the contents of these charts are very selective, with a mixture of abstract information like airways, and visible surface features which might be landmarks at operating altitudes or navigational hazards at low altitudes. These charts are certainly maps, but designed for users at several kilometres above the surface. There should be equally little dispute that a street atlas used for navigation at ground level is a map in everyone's terms. However, there are equally familiar forms of transport information that are less often thought of as maps. These include the signs posted at roundabouts and other complex intersections on the road network.

These three examples all have the same basic function. They are designed to help navigation. The examples are progressively more specialised and stylised. Aeronautical charts are navigational aids which combine precise navigational information with attempts to show landscape realistically from high latitudes. Street atlases have to serve a generalised community of users on the road or on foot, whose journeys might run between any pair of addresses. Because of this they have the appearance of traditional maps, covering the road network comprehensively and showing a large number of places of interest. Road intersection signs must provide clear instant information for users who cannot stop to study it. The result is a very simple and stylised map, designed within a national and increasingly international set of conventions. Sometimes the term 'diagram' is used to distinguish maps of this type from those which abide by conventional ways of showing distance and direction, but the boundary is arbitrary, and the term will not be used here.

Maps users should always be aware that no map tells the total truth. All cartography selects information to display on the basis of relevance, and many maps distort in ways which may not be obvious to the user.

Turning back to our examples, aeronautical charts ruthlessly exclude surface information which is irrelevant to navigation, but select and emphasise items which the ground user would not expect to find on the map. As a matter of course the designers of street atlases exaggerate the widths of roads enough to make room for the street names which are vital parts of the map. Transport route maps and signs cheerfully sacrifice scale for clarity. For example the familiar London Underground map has no consistent scale, but allocates most of its space to the dense detail in the central area. There is a seamless general change from centre to periphery. In current editions of the map, the distance between stations in central London can be as great as 31 map units per kilometre (London Bridge to

Bank), whereas in the outer areas distances shrink to as little as 1.6 map units per kilometre (Croxley to Watford). Road diagrams like the one in Figure 7:4 are gross simplifications, and always put the direction of travel at the top. If roundabout signs followed the convention of putting north at the top, the effects on the user could be disastrously confusing.

Professional map production

Most of the maps used by geographers are produced professionally by cartographers working within large organisations of different types. The examples we have just seen in the transport field were produced by international bodies (the aeronautical chart), commercial companies (street plans), service providers (the underground map) and government departments (road signage). All professional agencies produce maps under constraints of time and money. Some are straightforward commercial ventures in any case, but in the last few decades the tendency has been that even national mapping agencies have been under pressure to be commercially self-sufficient while still providing an effective service.

The Ordnance Survey provides an excellent example of a major national agency. It was founded in 1791, and since the 1840s it has been the primary mapping agency for England, Wales, and Scotland, publishing national topographic maps and specialist maps, and providing a projected map base used by other governmental agencies and most commercial publishers of maps of Britain. It publishes topographic series on a range of scales from 1:1,250 down to 1:1,000,000. For a clear history of the Ordnance Survey and a detailed review of its products that is still basically accurate, it is worth reading Harley (1975). Comprehensive coverage has never been possible. The Ordnance Survey's standard 1:10,000 map covers an area of 25 square kilometres, and complete coverage of this area at scales up to 1:1,250 would require an additional 125 sheets. On a national scale, this would require well over 1 million maps, many of which would be almost empty of information. Even in more spacious days, it was recognised that the effort of surveying and mapping upland Britain at the scales of 1:1,250 or 1:2,500 would not be justified by demand, and these scales do not extend beyond urban areas and the settled lowlands respectively. Even so, the Ordnance Survey is responsible for several hundred thousand maps, which have to be revised regularly and kept in stock so as to be available to the customer on demand. Revision is one of the major problems. The Ordnance Survey's working practice has been to revise in a hierarchical way, modifying key types of information like road networks ahead of other types of detail.

This creates 'time-transgressive' maps on which information of different ages sits side by side, presenting yet another problem from the user's point of view.

The development of computer-based mapping has revolutionised the way in which the agency works. In a process which started during the late 1970s and is now largely complete, the Ordnance Survey has digitised its operation, moving from printing plates to magnetic media as the main storage mode. The advantages of this are impressive. From the agency's viewpoint, its maps can be edited and revised much more rapidly than in the past. Users can be supplied under licence with maps in digital form suitable for use with particular types of software. Those users of large-scale plans who still want maps on paper can have copies printed on demand. This greatly reduces the storage problem for the supplier, and guarantees that the user gets the most recent edition. The professional user has the additional benefit of being able to request maps containing only certain types of information, and covering areas different from the traditional sheet lines. As a spin-off from the digitising process, the Ordnance Survey is able to offer special products like heighting information for the construction of three-dimensional models, and road centre-line data suitable for route-finding software and in-car navigational systems.

The Ordnance Survey is of course not the only map agency making extensive use of digital mapping. However it is a useful example of the ways in which new techniques are creating new services, not all of which were envisaged when the conversion was started.

The student as cartographer

So far we have looked at maps as the output of professional cartographers. They are important resources for geography students, and it is important to know what is available, and what their limitations are. Apart from the design issues we have already examined, maps are expensive, and sadly this is particularly true of many of the more recent digital products. Although most student users will have access to a teaching or research collection, it may not contain the maps needed for specific dissertation or research purposes. It is also quite possible that commercially produced maps are simply not available to meet a particular need. Inevitably there will be occasions when it is necessary to produce original maps rather than use existing products. Most geography students at one time or other will find themselves preparing maps either as tools for use in analysis or as elements of presentations.

Over the last thirty years the technical resources available for student cartography have changed tremendously, starting with drawing ink and stencils on paper, moving through a phase of drawing and shading film and prepared lettering, to the present situation in which many students have access to mapping software. The speed of change means that up-to-date technical manuals are not available, and this is not the place for a guide to any one of these ranges of techniques, although we will look briefly at computer mapping software at a later stage. Here, it is much more important to consider issues that relate to the conception and design of maps for presentation.

In general, fitness for scale and purpose should be the guide. This almost totally eliminates the use of photocopies of existing maps. Leaving aside the danger of copyright infringement, which is almost always present when maps are photocopied, it would be very unusual for an existing map to carry no more and no less than the information required for a specific project. To ensure that the map is entirely appropriate, it is really necessary to do it oneself.

This means taking charge of the map from the design stage, which is one of the few areas of cartography which remains unautomated. No matter which mapping media are being used, it is the student cartographer's task to produce a map that is a visually attractive, clear and unambiguous presentation of the information and images that are to be transmitted to the intended audience. Robinson *et al.* (1994) identify the key elements in graphic design. Clarity and legibility are paramount. The map should provide an immediate visual impression, but also yield clear patterns when studied in detail. Visual contrast is important, and careful choice of line type and thickness, shading, and lettering will reinforce the impression of clarity. Balance is also essential. Most maps will contain a number of distinct elements. Apart from the body of the map itself, there will be a title and perhaps subtitles, a key or legend explaining the symbols used, and supplementary elements like scale-bar, north-point, and perhaps a description of the source of the data used. These elements should be set out in a way that is as harmonious and uncluttered as possible.

Robinson *et al.* go on to discuss other design issues, and it is suggested here that one of the more useful ways of moving on from this text is to read through the later chapters of their text, ideally with a map design project in hand that can be used as a test bed.

Summary

The point that underlies all the topics we have looked at here is that maps are documents. Like printed documents they have been written with purposes in mind, and the data used to create them has been selected for this purpose. Whether you are creating a map for your own needs or looking at one designed by a professional cartographer, the map can be read in two ways. Not only should you interpret the patterns it depicts, but you should as far as possible read between the lines. Although you will not normally know in detail how it was done unless you have drawn the map yourself, the choice of patterns to be shown or excluded is a vital part of the creation of a map.

Discussion questions and tasks

1 Select and deconstruct a topographic map on a scale of 1:25,000 or 1:50,000. What information is included and what is not? What sort of symbolisation is used and what types of information is it applied to?

2 Find a map of the same area on a smaller scale – 1:250,000 or 1:625,000 for example. What are the major differences?

3 Explore the margins of mapping. Look at maps designed for quick use by non-specialists – for example supermarket floor plans, theatre or aeroplane seating plans, street maps of tourist resorts. Are there common features among the ones that do their job more successfully?

Further reading

See also

Computer Mapping, Chapter 9

General further reading

Useful texts combining technical guidance with discussion of the aims of cartography include:

Keates, J. (1989) *Cartographic Design and Production*, Harlow: Longman, 2nd edition.

Keates, J. (1996) *Understanding Maps*, Harlow: Longman, 2nd edition.

Kraak, M.J. and Ormeling, F. J. (1996) *Cartography – Visualization of Spatial Data*, Harlow: Longman.

MacEachran, A.M. and Taylor, D.R.F. (eds) (1994) *Visualization in Modern Cartography*, Oxford: Pergamon.

Robinson, A., Morrison, J., Muercke, P.C., Guptill, S.C., and Kimerling, A.J. (1994) *Elements of Cartography*, New York: Wiley, 6th edition.

If you are enthusiastic you might explore this multi-volume set:

Anson, R.W. and Ormeling, F.J. (1988–1996) *Basic Cartography for Students and Technicians*, Oxford: Heinemann, 3 volumes and Exercise Manual.

Two texts that provide stimulating reading about the way that the purposes of cartography have developed are:

Wood, D. (1993) *The Power of Maps*, London: Routledge.

Barber, P. and Board, C. (1993) *Tales from the Map Room*, London: BBC.

Notes

[1] Peters, A. (1989) *Peters' Atlas of the World*, Harlow: Longman.

8 Using computer applications

The development of generic computer techniques means that all students are now expected to be able to make use of them. This chapter covers:

- An introduction to the machine
- Word-processing
- Spreadsheets
- Presentation graphics
- Computers in the field

The spread of computers in school-level education and homes means that many students now arrive at university thoroughly familiar with machines and software. However, every cohort still has members without this advantage, typically mature students whose school education finished before the late 1980s. This chapter is dedicated particularly to these non-initiates in the field of computing. Seasoned users should feel free to by-pass it, but might still find it helpful to browse.

Most institutions base their work on one of a handful of machine types or 'platforms', of which IBM-type machines and the Apple Macintosh are the most common. Even a few years ago the choice of platform was very important in determining the software available to the user, but recently there has been an impressive amount of convergence in the appearance and behaviour of software on different platforms. The examples in this chapter will be drawn from the IBM field, but readers with access to other platforms should still find the examples helpful. Because of limitations of space, the text cannot cover absolute basics like keyboard familiarisation and disk handling. Most universities provide their own in-house introductory guides, and you should make use of these.

This chapter provides an introduction to what are sometimes called 'generic' areas of Information Technology, the range of applications which are not specifically designed for workers in a particular field of study but will still be very important in enabling work to be done to a high standard. A proficient user not only benefits from the efficiency and analytic power offered by these applications, but can produce work that looks attractive. Despite the ingrained suspicion in academic circles that good presentation must somehow have been achieved at the cost of content, generic software makes it much easier to produce work that scores well on both counts. The fields covered will be word-processing, spreadsheets, and graphics packages.

This choice of applications is based on the relative friendliness of these applications, particularly word-processors and spreadsheets. A case could certainly be made for including their more sophisticated relatives, desk-top publishing packages and database managers respectively. However, both of these are notoriously difficult to use effectively without extensive training and experience. Word-processing and spreadsheets give much more immediate access to a range of important facilities and are also platforms from which the user can step, when the time is right, in the direction of the more powerful packages.

Because there are so many competing applications in these areas, it would not be sensible to provide a manual describing the operation of particular pieces of software in detail. Instead, it is intended to demonstrate how these applications can be used to get the maximum value from their features. At the end of the chapter the reader should have a clear overview of the benefits provided by these applications, and be aware of the ways in which they are typically organised. He or she can then go on to learn how to make a particular application perform.

An introduction to the machine

Most new users in academic life will first meet a personal computer or PC. Although large centralised mainframe machines and high-powered workstations may be used for specialised purposes, the trend through the 1980s was for routine work to be transferred to stand-alone PCs. The 1990s have seen a shift towards the networking of PCs, with centralised servers storing software for groups of PCs and co-ordinating their demands for resources like printers. This should not really matter to most users. In the terminology of computing the change has been 'transparent', so that the user sitting at the keyboard does not need to know whether or not the machine is part of a network. Another important development of the early 1990s was the appearance of genuinely cheap and powerful portable machines. We will look at them later in the context of fieldwork.

Improvement in machine performance has been striking, and has not levelled off. There is little point in comparing the technical specifications of the machines used even five years ago with the machines that students will typically be able to use at present. Readers who understand the technology will already know something about the scale of the change, and those who are new to the field will not be enlightened by a string of technical details about speed, memory, and storage. Most students can now reasonably expect to be able to use fast machines with high-resolution colour screens, supporting Graphic User Interfaces (GUI) like Microsoft Windows 3.1 or Windows 95 designed to make life easier for the user. The computer will probably have external connections for electronic mail (email) and Internet access, and it will probably also have sole or shared access to a CD-ROM reader and specialised input devices in the form of scanners and digitisers. Sharp reductions in prices mean that high-quality inkjet printers are now common, and networked PCs might offer their users access to a laser printer or plotter as well. A user with a few years experience might notice just one area where there has been little progress. Although screen resolution has improved greatly large monitor screens remain expensive, and student machines still usually have screens no larger than 14".

Most student users will store their own work on diskettes (floppy disks). The larger 5.25" format is now more or less a museum piece, and most machines accept only the more robust 3.5" disk format. The standard high-density (HD) 3.5" disk has a storage capacity of 1.44 MB, sufficient to store a year's essays and projects (or the text of a short book like this!) comfortably. On the other hand disks are cheap and it is a false economy to try to cram too much on to a single disk. Before their first use, all disks have to go through a process called formatting, which allocates the storage space in a way that is compatible with the type of machine you are using. Ready-formatted disks are now easy to get, but if you have to format your own, follow the instructions provided by your institution or machine manual.

Disks are fairly tolerant of bad treatment, but do not deserve it. They should not be used as coasters or left on sunny window-sills, and they will not survive immersion in pools of coffee or soft drinks. They should not be stored on top of equipment that sets up magnetic fields, like television sets, speakers, or even the monitor of the computer itself. Some users are concerned about exposure to the powerful fields set up by the underfloor motors of underground and electric surface trains, but this author has logged thousands of disk-hours of underground travel without any evident problems. Nevertheless, things can go wrong, and one of the best safeguards against accidental damage or corruption is to back up any files worth keeping. A second copy of any important file should be made on a separate disk, which should then ideally be stored in a different place. There has been a great deal of worry in the last few years about viruses. Most universities now have routine screening systems which will check all disks inserted by users or halt a machine if a virus is detected, and this has certainly limited their spread. However, it is still possible for them to find their way onto the system, usually because users with home computers have picked them up by borrowing software from infected sources, then bringing their own disks in complete with the infection. Machines behaving oddly should be reported to the local computer staff. Even if they are not playing host to viruses they probably need attention in any case.

One of the features of the later 1990s is going to be a revolution in portable storage. In 1995 two high-capacity drive systems were launched which allow the user to store up to 135 MB on a small-format cartridge. The cartridges are more or less affordable on student incomes, but obviously need host machines with suitable drives. Competing products under development in 1996 will offer as much as 1 GB in a 3.5" format, and in the slightly longer term the CD-E (CD-erasable) will offer even greater storage potential. Storage on this scale will far exceed any need you might have for text, but may be very important if you have to store big graphics files relating to the applications discussed in later chapters.

Word-processing

The term *word-processing* is now part of the language, although it no longer comes close to describing the range of things that can be achieved with word-processing software. Modern word-processing software almost universally works within the spirit of the familiar WYSIWYG philosophy – What You See Is What You Get – and what you get is much more effective than the software available only a few years ago. The earliest word-processing was more or less the on-screen equivalent of traditional typing. It differed from the primitive text line editors of the early days of open-access computing mainly in offering a view ranging from a few lines at a time to a whole page. Later forms often had clumsy embedded commands which made it difficult for the inexperienced user to relate the screen display to the final output. With current word-processing software, the user always has a clear image of the way the text will appear when printed. The description that follows is based on the features available in Microsoft Word for Windows, but most other major word-processors share the main features described.

The most fundamental advantage of the word-processor over the typewriter is the ability to rebuild the text. The only clean way of amending a typed or hand-written document is to add new text at the end. However with a word-processor elements of all sizes from a single character to a whole file can be modified no matter where they occur. Positioning a screen cursor allows text to be typed in or deleted at any point in an existing document. The new text will adjust to take account of the changes. At a slightly more advanced level, blocks of text of any size can be 'marked' for treatment. This treatment might mean that the marked block is cut out entirely, moved to a new location, or duplicated elsewhere in the file. Environments like Microsoft Windows offer a further refinement. A marked block of text can be copied to the so-called 'clipboard', then later 'pasted' into another word-processed document or even a different application entirely. It is thus possible to import and export text between different files, whether they belong to the original author or a different person. Word-processed text is dynamic and mobile.

A second key benefit of word-processing is that text can be defined in modules which can then be given special treatment. On the traditional typewriter, the user marked the end of the line by using the 'Carriage Return' apparatus, which mechanically moved the action to the beginning of a new line. Although the term 'Return' is still widely used as an alternative name for the 'Enter' key on the computer keyboard, its use is not at all like the Carriage Return of the typewriter. By default, it is normal for word-processors to allow text to 'wrap around'. What this means is that as the user carries on typing in text, the software works out a suitable position in the text to move to a new line and automatically moves on to the next without the user's intervention. As long as the Return key is not used this carries on indefinitely. However using Return forces the software to move to the next line, and also adds an invisible marker which defines the end of a section of text, or 'paragraph' as it is usually known. In other words you use the Return key to mark the end of paragraphs rather than lines. The significance of this is much greater than it might seem. The ends of lines within a paragraph have no fixed identity. If text is added to a paragraph or deleted, the whole thing automatically adjusts to compensate in a way that line-defined text could not. Furthermore, whole paragraphs can be treated as units when it comes to formatting, which we shall look at next.

Flexible formatting is a third major advantage of word-processing. Some formatting instructions can be applied at any scale from the individual character to the whole document. These include the definition of font or typeface, character size, and refinements like the use of bold or italic styles. Other formatting instructions apply to the paragraph. Two of the most important are indenting and alignment. Indenting is width of the paragraph in relation to the working area on the page. Whole paragraphs can be indented, or special arrangements made for their first lines. Alignment determines how the text behaves when it wraps around. Most users' preference is for the text to be left-aligned (with the text flush to the left margin but ragged on the right), or justified (with the spacing automatically adjusted so that both left and right margins are flush). Headings are sometimes centred on the page. All of these actions can of course be modified or cancelled according to taste.

Page formatting allows the overall size of the working page to be changed, and it also allows the user to build in special features. These obviously include page numbers, as well as headers and footers (the continuity lines sometimes used in documents to provide an identity on every page). Footnotes or endnotes are also page-related. Most word-processors will be able to maintain a series of numbered notes and re-label them in response to insertions or deletions, as well as providing an automatic solution if a footnote expands to the stage that text has to be displaced to the next page.

All of these features can be defined as the need arises, but one of the more powerful developments of word-processors is the user's ability to build customised specifications which can be called up repeatedly. Sets of formatting instructions are usually called 'styles'. Most word-processors come with a series of styles. There will be a default style that defines character and paragraph features whenever a document is opened. There will also be several additional styles defining a range of headings and sub-headings, and the user will be able to build and save personal styles which can be applied whenever appropriate. Macros are more powerful. A macro is a series of linked commands provided by the package or built by the user, and accessed by using a keyword. Typical macros might search the text for particular pieces of text and change them, ensure that files are saved to a particular place, or set up and carry out printing instructions. As a warning, it has to be said that the ability to create styles and macros may be very restricted on networked and other common-user machines, if the administrator responsible for controlling the machines wants to control users' freedom to modify the working environment.

Word-processors also come equipped with tools for proof-reading and preparing documents for final output. Some of these are simple but very useful. It is usually possible to obtain a word count for the document, an important factor if working within length limits. There will be facilities for finding particular words or phrases where they occur in a document, and replacing them if necessary. One of the most important features, although surprisingly often neglected, is the spelling checker, which allows the user to verify spellings as appropriate, and make whatever changes are necessary. Most applications also provide a thesaurus (allowing the user to substitute a synonym for an over-used word) and a grammar-checker. For more ambitious projects, it is possible to use revision marks to track changes made in the document, and to mark text in particular ways in order to produce an index or table of contents.

Word-processors normally allow the user to insert information outside normal text forms. The most straightforward is the table. The user can define a table as a matrix of cells that provide a more disciplined framework for the display of numerical values or text than the open page provides. The great advantage of using a formally defined table is that it allows the user to control more effectively not just the characteristics of individual cells but also the rows or columns that form the table as a whole. Tables can also be used as host environments for cells imported from spreadsheets, or graphics images of the type we will meet later.

One of the most exciting advances in software in recent years has been the development of objects, semi-autonomous entities with their own characteristics that can be transferred between different applications. Many applications supported by Microsoft Windows can exchange objects freely. Microsoft provides a series of small applications or 'applets' that allow the user to 'embed' objects into a document created in Microsoft Word, then return later if necessary to edit or modify the object. These give the user the power to add a wide range of extra features, including drawings and graphic images, pre-designed artwork (known as 'clip-art'), and even equations. Graphs can be created using data taken from tables already existing in the document or tabulated in the embedded application itself. Graphic objects created in some other Microsoft Windows applications can be embedded in the same way. A refinement of this is the use of links. Linking means that the object is not only copied into a word-processor document, but retains contact with the original source, so that if the original should happen to be modified at a later stage, the user of the word-processor document will be able to update the linked copy to reflect the changes. Storing the linked documents on the same diskette allows the link to be preserved in a common-user system.

All in all, word-processing gives the user fast and powerful control over the way that documents are developed. The example in Figure 8:1 shows this. The finished text in Figure 8:1b was derived from the raw text in Figure 8:1a, and two pre-defined objects were copied over from a spreadsheet and a graphics application respectively. The whole process took approximately three minutes, and could have been changed radically again in another three. It is not surprising that people who have learned how to make the most of word-processing seldom want to return to hand-writing.

(a)

Environmental damage and attempts to control it

The early Leblanc alkali process was not an environmental asset. Clouds of acrid fumes emerged from the works every few hours, and each works had a large unpleasant-smelling waste heap. James Muspratt, one of the pioneers in Lancashire, continued to run a number of works after the 1820s, but persistent legal pressure initiated by Liverpool Corporation and landowners because of fume emission closed his works in Liverpool and Newton-le-Willows in 1849. However he was able to retain later works which had been built to rather higher standards.

These were not the first complaints about the industry. This is hardly surprising, since it emitted large quantities of hydrochloric acid gas – estimated to be as much as 60,000 tons per year from the main districts in the 1830s. Complaints on Tyneside in 1839 emphasised the death of vegetation, crops, and livestock, the tarnishing of bright work on domestic furniture, and human ill-health.

Some of the attempts by the chemical companies to defend themselves were almost comical. In 1843 when a firm in South Shields was being attacked for its gas emissions, the workers held a 'spontaneous' grand parade to the town hall to defend their employers. A particularly large and healthy-looking employee was exhibited publicly as evidence of the beneficial effect of the atmosphere in the works. An industrialist addressing a meeting at Jarrow around the same time said

'Gentlemen, some people say the manufactories of this borough are injurious to health. I don't believe it: The healthy faces of everyone around me prove it false. Look at ourselves – Why, this is one of the healthiest towns in all Her Majesty's dominions.' 12

In 1860 an alkali manufacturer wrote to Chemical News to say that one of his foremen had once had a lodger suffering from tuberculosis. This man, visiting the works, accidentally breathed in the hydrochloric acid gas. The effects were so beneficial that he started visiting the works regularly to breathe the gas as a form of health cure, and eventually became so robust and healthy that the manufacturer gave him a job. Whether the writer genuinely believed this or not, alkali works did not replace Alpine sanatoria as resorts for tuberculosis sufferers.

12 Quoted by W. A. Campbell in The Chemical Industry (1971:37)

Figure 8:1 *Word-processing: before and after*

(b)

Environmental damage and attempts to control it

The early Leblanc alkali process was not an environmental asset. Clouds of acrid fumes emerged from the works every few hours, and each works had a large unpleasant-smelling waste heap. James Muspratt, one of the pioneers in Lancashire, continued to run a number of works after the 1820s, but persistent legal pressure initiated by Liverpool Corporation and landowners because of fume emission closed his works in Liverpool and Newton-le-Willows in 1849. However he was able to retain later works which had been built to rather higher standards.

These were not the first complaints about the industry. This is hardly surprising, since it emitted large quantities of hydrochloric acid gas – estimated to be as much as 60,000 tons per year from the main districts in the 1830s. Complaints on Tyneside in 1839 emphasised the death of vegetation, crops, and livestock, the tarnishing of bright work on domestic furniture, and human ill-health.

Key dates in the Leblanc Process	
1825	Lussac's Tower
1836	Gossage's Tower
1839	Pyrites as base for sulphuric acid
1853	Revolving Furnace
1858	Henderson's pyrites-copper process

To the right, a typical alkali tower

Some of the attempts by the chemical companies to defend themselves were almost comical. In 1843 when a firm in South Shields was being attacked for its gas emissions, the workers held a 'spontaneous' grand parade to the town hall to defend their employers. A particularly large and healthy-looking employee was exhibited publicly as evidence of the beneficial effect of the atmosphere in the works. An industrialist addressing a meeting at Jarrow around the same time said

> 'Gentlemen, some people say the manufactories of this borough are injurious to health. I don't believe it: The healthy faces of everyone around me prove it false. Look at ourselves – Why, this is one of the healthiest towns in all Her Majesty's dominions.'[12]

In 1860 an alkali manufacturer wrote to **Chemical News** to say that one of his foremen had once had a lodger suffering from tuberculosis. This man, visiting the works, accidentally breathed in the hydrochloric acid gas. The effects were so beneficial that he started visiting the works regularly to breathe the gas as a form of health cure, and eventually became so robust and healthy that the manufacturer gave him a job. Whether the writer genuinely believed this or not, alkali works did not replace Alpine sanatoria as resorts for tuberculosis sufferers.

---

[12] Quoted by **W. A. Campbell** in *The Chemical Industry* (1971:37)

However it has to be said that traditional techniques have their advantages. Some users find that word-processing is tiring except in fairly short sessions. Many others find that a screen display is not ideal for proof-reading. Although spell checking solves some problems it is not uncommon to spot errors in logic and grammar immediately from a printout that were not obvious while the file was on the screen. It is a good idea to make a hard copy and read it carefully before submitting the final draft of a word-processed essay or project.

Spreadsheets

Spreadsheets are perhaps not as familiar to geographers as word-processors, but they should be. Current market leaders in this field, like Microsoft Excel, Lotus 1–2–3 or Novell Quattro for Windows, have a range of facilities as wide in their distinct way as those of word-processors. The example used here is Quattro, but the leading spreadsheets all have very similar facilities, and it should not be assumed that any of the features described below are exclusive to Quattro.

The immediate impression given by a spreadsheet is that it presents large amounts of data in orderly and visible rows and columns. However the real power of a spreadsheet lies in the ability to make the data in different locations or 'cells' communicate with each other. Spreadsheets were first developed for the commercial world, in order to manage and present budgets and other financial operations. In this context the ability to combine fluidity of operation with automatic update of totals and other summary figures was invaluable. Because text labels were also essential in contexts like these, spreadsheets were designed to cope with text handling as well as numerical operations, and this made them suitable for inventory and catalogue work. It also makes them very suitable for our purposes.

Slightly more formally, a spreadsheet can be described as an application in which data can be stored in a structured way, so that relationships between different data items can be established. Since a quick definition like this could serve equally well to define a formal database management system (DBMS), it might be best to clarify the difference between a spreadsheet and a DBMS straight away. A spreadsheet presents a visible matrix of data cells, each one of which has a unique address defined by its column and row (X and Y) location within a 'page'. Figure 8:2 shows part of a typical page, a brief analysis of UK road traffic accidents. When a new spreadsheet is opened all the cells are empty or latent, and the user fills only those necessary for a particular purpose. An occupied cell will

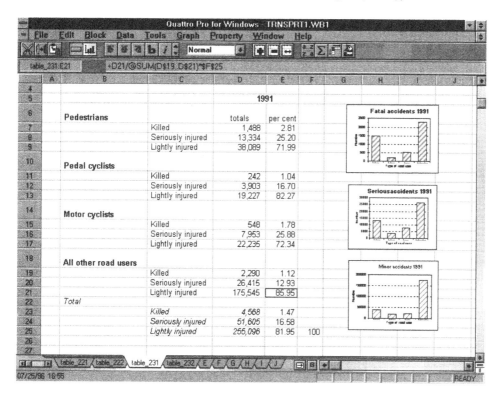

Figure 8:2 *A screen image of a spreadsheet*

probably be part of a spatial series containing similar data, and might be linked by formulae to cells at other addresses on the page. The visible and co-ordinate-based organisation of a spreadsheet makes it in the broadest sense 'geographical'. In a DBMS, on the other hand, information will normally be organised into sets of 'records', each record containing data in different 'fields', and the data type and storage size permitted in each field must be defined in advance. In a DBMS the record is the primary unit of organisation and although a set of records can be displayed on the screen there is no counterpart to the co-ordinate system of the spreadsheet. There is a rigidity of discipline and hierarchical organisation (fields within records), that does not constrain spreadsheets.

Although some introductory texts on spreadsheets recommend that the user should think of the rows as database-type records and the columns as the fields belonging to them, you should not let yourself be restricted in this way. Cells with quite different contents can be interleaved, and the data type within an individual cell can be changed without difficulty. The only fundamental restriction is that the data in a cell must belong to a single one of the dozen or so numerical and text types available. The storage potential of a spreadsheet is enormous. In Quattro, for example, each page contains 256 columns and 8,192 rows. Since a Quattro file or 'notebook' has 256 pages, the theoretical total in any file is well over 500 million usable cells, although you will never need more than a tiny fraction of this.

From the point of view of the human geographer, the most common use of a spreadsheet will be to build tables which need some kind of numerical analysis. No other form of generic application is as well adapted for this purpose. What is actually done with the data will depend on the user's requirements and the scale of operation. Spreadsheets provide a very fast and neat way of producing small tables which can be imported into word-processors to support the text. In many other cases the analysis will be carried out entirely in the spreadsheet, and parts of the spreadsheet can be printed out or used to produce formal output reports. As we will see later, most spreadsheets can also produce good portable graphics. If statistical processing is required beyond the range supported by the spreadsheet, data can quite easily be exported to more powerful specialised applications such as SPSS for Windows.

Spreadsheets can also be used for text handling, but here the advantages are not so clear-cut. Nobody would prefer a spreadsheet to a word-processor for producing conventional documents, and if text simply has to be tabulated neatly, the Table function of a word-processor is likely to be at least as effective. Nevertheless, there are forms of text analysis where the spreadsheet should be considered. Many spreadsheets, like Quattro, let the user query text as well as numerical data in the way that is typical of databases, but without the overheads of setting up a formal database and learning one of the rather forbidding query languages used by DBMS. Suppose that a retail geographer is analysing the results of a questionnaire of shopper preferences. By setting out the results on a spreadsheet page, it is not difficult for the researcher to carry out rapid queries to assess the proportion of respondents who have (for example) assigned high- or low-score labels to a shopping centre on different criteria. Even cells containing unstructured lists of key words obtained from individual respondents - 'clean', 'cheap', 'crowded', and so on – can be searched to analyse the pattern of occurrence of particular words. Fairly complex queries are possible. In the example sketched out above, for example, it would not be difficult to establish the proportion of shoppers who rated a centre highly in most ways but had expressed reservations in some respects.

At this point it might be helpful to provide a little more information about the ways in which spreadsheets link cells to each other. Some of the occupied cells on the page, usually the majority, will contain raw data. The linkage is carried out by typing or copying into individual cells functions and formulae which include the co-ordinate addresses of cells which already contain data or other formulae. In this way the user can establish a dynamic link between cells, so that the contents of specified locations will consistently be updated if changes take place at other addresses. To give a simple example, a cell designated to contain the sum of a series of other cells will automatically display an updated total if any one of the series is changed. In the spirit of WYSIWYG, the visible display for a cell containing a formula will always show the outcome rather than the formula itself. This can be seen in Figure 8:2. The highlighted cell (E21) displays the numeric value 85.95, but the input line above the page shows that this is derived from a percentage formula making use of values in D19, D20, D21, and F25. When a cell is highlighted like this, its contents can easily be edited or replaced. In fact the lack of warning about data loss when working in a spreadsheet can be a problem, and slapdash data entry might be heavily punished.

Most spreadsheets provide a large menu of functions. Quattro, for example, has about 120 functions for statistical, trigonometric, logarithmic, and other purposes. The user can also define simpler or more complex formulae using numbers, cell addresses, and functions linked by arithmetical or logical operators as appropriate. The only limit to this is appropriateness. The pre-designed statistical functions in Quattro, for example, go only as far as standard deviation and variance (see Chapter 5), although more advanced techniques suited to the cell-based organisation of spreadsheets, like regression analysis and matrix arithmetic, are also available. User-defined formulae could certainly be used to perform more advanced statistics, but if possible it is better to carry out statistical work using specialised software like Minitab or SPSS for Windows, where the tests are proven and robust, and useful features like significance levels are available on tap. It is not difficult to import spreadsheet files into SPSS for Windows.

A lot of operations involve repetitive application of the same formulae to different cells, and spreadsheet designers have been quite ingenious in accelerating the way that this can be done. At the risk of becoming unduly technical it is worth explaining the ways in which spreadsheets use relative and absolute location. An understanding of this will emphasise the power available to spreadsheet users. Relative location is the default mode in Windows spreadsheets like Quattro and Excel. The logic of relative location is that if a formula or function is copied or moved to a new location, the addresses in the formula are automatically converted so that the copy looks for data in the same relative positions.

If a formula in cell I8 which reads data from its neighbour two cells left (G8), is copied to K16, the new version will look for data in the equivalent neighbour two cells left (I16). Relative location means that a formula can be copied instantly to whole blocks of cells. In other words a few seconds of work with the mouse can apply the same formula to tens or hundreds of different groups of data. There are other times when it is important for an absolute location to be defined, in terms of rows, columns, or both. In Figure 8:2 the highlighted formula (+D21/@SUM(D$19..D$21)*F25) underlying the value in E21 shows absolute location in action. Quattro uses the dollar sign to mark absolute row or column locations, and in this case F25 refers to the percentage multiplier located at the intersection F25. All the other live cells in column E use versions of this formula and refer to F25, and no matter where we copy this formula, it will go to F25 for the multiplier. If we were to change the value there, all cells using it would modify their contents accordingly. Absolute location allows us to define complex formulae just once, or force widespread change on the basis of alteration to one cell.

Another of the strengths of the better spreadsheets is the ability to produce high-quality graphics from their contents. In Quattro there are over twenty two- and three-dimensional bar, column, line, or pie types available, and it is also possible to rotate these or incorporate multiple copies in the graph. The graph itself is created by interactively defining different blocks of cells to provide the graph with data and labels, and changing the visual properties of the graphic as appropriate. This takes place in a drawing environment which allows the graph to be enhanced with additional linework and text and a wide range of alternative shadings. The process is intended to be interactive, quick, and flexible. On the whole it succeeds, although it lacks the polish of the presentation graphics examined below. Quattro graphs are automatically updated if their component data are altered, and automatically saved as part of the notebook. As with other Windows objects, they can be exported to word-processors or other applications as embedded or linked objects of the types described earlier. An entertaining and sometimes useful feature within Quattro is the ability to insert a graph into the data page, presenting a visual image of the data which will be visibly updated if the data are modified. Figure 8:2 includes thumbnail graphs alongside the data to which they refer. Objects from other Windows applications, even video clips, can be embedded in spreadsheet pages too, although this may have more novelty value than usefulness.

Presentation graphics

So far we have seen that both word-processors and spreadsheets routinely allow the creation and copying of graphics to quite reasonable standards. However there is also specialised software generically known as presentation graphics. As the name indicates, these applications are designed to support verbal presentations. The end-product will typically be a 'slide-show' of images in sequence using text, drawings and clip-art, and graphs. Well-known and easily accessible applications in this field include Harvard Graphics, Persuasion, and Lotus Freelance.

Do presentation graphics like these offer advantages that make it worth while learning about yet another application? As always, the answer to this question depends on what we want. We have seen that word-processors and spreadsheets already give us access to a wide range of techniques, and they would certainly allow us to produce competent graphic support for essays, reports, and dissertation work. Having said this, there are probably two advantages of learning a presentation package as well. The first is that it would provide a tool designed to support seminar papers and other student presentations in a clear and attractive way. Although word-processors can do this, presentation graphics undoubtedly do it better, and more easily as well once the techniques have been mastered. Much the same is true of special effects. Presentation graphics provide a range of image processing techniques that the non-specialist tools we have already looked at cannot match. Whether advanced image treatment would be appreciated in the rather staid context of an academic presentation is another matter!

Computers in the field

One of the most exciting developments in computing in the 1990s has been the development of affordable portable or notebook machines with good specifications. The prices of mobile machines have fallen steeply, and in the mid-1990s it is possible to buy portables cheaper and with much better specifications than their desktop counterparts of the late 1980s. From the geographer's point of view one of the most interesting aspects of this change is the possibility of taking the machine into the field. The advantages of by-passing the logging sheet or questionnaire form stage by keying field data directly into its final storage form and even doing provisional analysis in the field do not need to be spelled out. There are no technical difficulties in designing a questionnaire which respondents could answer interactively in the street using a screen pointer.

However, the portables of the current generation are not ideal for direct field use. Although conventional notebooks are much more robust than earlier machines, genuinely all-weather machines like the data loggers issued by public utilities to their field data recorders are still expensive. Working sessions in the field are usually rather short. A typical portable manages between three and four hours of working life per battery charge. Some smaller 'sub-notebook' machines have longer session lives, but the power is saved by offering a lower specification, including screens which are less visible in outdoor conditions. The field sites typically frequented by human geographers often have their own special problems. Streets are not well-equipped with flat level working surfaces, and it is not practicable to use a full-size portable as a hand-held machine. Sub-notebooks can be used hand-held, but do not leave a spare hand for manipulating paper, and it is difficult to operate their miniature keyboards with numb fingers on a cold day. There are also security problems. Some outdoor urban work exposes the machine to a serious risk of theft, and all the stored data would inevitably disappear along with the computer unless you were able to back up regularly.

Despite these problems, fieldwork can be done with a portable. If it is possible to work as a group or use a vehicle as a base, the difficulties become much less pressing. At a future date improvement in voice recognition techniques might revolutionise the use of machines in field data collection. As long as the keyboard is the main mode of input to the portable machine, though, it is probably best to leave it in a safe place until the data collection has been done. It can then be used effectively for data input and analysis at the end of the working day. A good portable will be able to support all the software needed for field purposes, and a modem link will also open up access to software and data stored at a university or other remote site.

Summary

A brief survey of generic software does not provide the scope for doing much more than touching on some areas that might be useful, and others, like time planning and project organisation, have had to be disregarded entirely. The applications we have surveyed all provide an introduction to others which are too specialised for this text. The facilities offered by word-processors offer only part of the text manipulation power of desk-top publishing (DTP). Spreadsheets, as we saw earlier, cannot provide the rigour and depth of analysis offered by database management systems. Basic presentation graphics provide only an introduction to the range of specialised graphics and image treatment applications.

One important development in the last few years has been the expansion of multimedia. Fast processors, the storage power of CD-ROM, and the linkage between applications built into front-ends like the evolving forms of Microsoft Windows have all encouraged this. As applications in specialised fields have become more advanced, there has been a tendency for their more basic features to be incorporated into ordinary generic applications. Word for Windows, for example, has multimedia links which allow documents to incorporate not only graphs and images but audio and video clips. Spreadsheets too have developed multimedia capabilities, although the advantages here are not so obvious. Learning to use one or two basic generic applications now gives one access to a whole range of computing facilities previously available only to specialists, using multimedia links that even specialists did not have a few years ago. The pressure of competition means that as time goes on, the developers of basic applications will continue to offer extra benefits. For example mapping facilities have been available with some spreadsheets since 1994, although they are still based on libraries of pre-defined maps. It may be that applications will become even bigger and clumsier, and some of the new features may have little practical value, but on the whole this trend will be to the user's benefit.

Discussion questions and tasks

1 Be positive and creative. Take a handwritten essay, plan a better presentation, and see how far you can achieve this with a word-processor.

2 Explore portability. Find out what you can do by linking and copying between word-processing, spreadsheet, graphics, and other software.

3 Think about the difference between absolute and relative addressing in spreadsheets. Why is it so important?

4 Work out a back-up strategy before you find out the painful way why it is necessary!

Further reading

See also

Statistics, Chapters 5 and 6
Mapping techniques, Chapter 7
Computer mapping, Chapter 9

General further reading

It is not easy to recommend further reading in this field. Texts about computing in geography ignore the 'Cinderella' subject of generic software, but it is also surprisingly weakly handled in skills texts like *The Good Study Guide* (Northedge 1990). There is no shortage of commercially published texts about these techniques, but they almost all deal with individual commercial products, and the choice available to you will depend on the software you are likely to use. Computing texts tend to be rather expensive. In all cases find out whether your university produces its own software guides before buying a textbook.

⬤9 Computer mapping

Slowly but surely computer applications are being produced which allow geographers to handle spatial data on the screen. This chapter examines:

- Storing and displaying spatial images
- Computer atlases
- Do-it-yourself computer maps
- Using mapping software
- Computer Aided Design and AutoCAD
- MapInfo

In this chapter we move from generic to specific applications. No technique is 100 per cent geographical, and spatial techniques are not the exclusive property of geographers, but there is a group of map handling techniques that are entirely appropriate here. The criterion is that they should be geographical in the sense that they deal with spatial patterns and relationships, and are designed particularly to deal with questions that geographers tend to ask.

We will start by looking at the ways in which maps are put on the screen, and the range of ways in which computer mapping can be useful to the student. After sketching out the basis of map display on the screen, we will look at the range of pre-designed maps available. The next step is to look at ways of mapping patterns for oneself. Finally we will turn to the way in which particular packages can be put to use in geographical contexts.

This is not an area well served by the existing literature. Although a number of books have been written about computing in geography, they have tended to emphasise the physical side of the subject. They have also tended to become outdated very fast as techniques have changed, a fate that faces any book about computing, this one included.

Storing and displaying spatial images

Before we look in detail at different techniques, it is necessary to look briefly at the way that the machine handles spatial information. The most important point relates to the way in which images are built up, displayed, and stored. The screen pattern you see at normal viewing range is actually composed of a great many individual cells or 'pixels', so small that it is not possible to pick out individuals with the naked eye. Under magnification these pixels would appear as horizontal lines of dots repeating the sequence blue–red–green, with successive lines offset so that the three colours do not appear as vertical stripes. Even when the image displayed on the screen seems static to the viewer, the machine is constantly retrieving it from a store known as a 'bit map' and scanning it into the pixels on the screen, which happens at least sixty times per second to avoid visible flickering. This pixel-by-pixel, line-by-line projection on the screen is a 'raster scan'. The resolution or level of detail the screen can display is determined by the number of lines of pixels displayed in a given area. There is a bewildering range of standards of resolution, but those commonly used for Windows applications are VGA (640 × 480 pixels, or 307,200 in total) and Super VGA (1,024 × 768, or 786,432). A range of colours is also made available by varying the tone and intensity of the three basic colours in pixel combinations that can define anything between sixteen and several million colours, depending on the graphics mode and the software in use.

We do not always need this richness of information, and machines can run faster at lower levels of resolution. The conventional text mode dates back to the time when they had very limited capacity. It simply divides the screen into a matrix of 25 lines of 80 character or letter positions (2,000 in all). Characters are formed by assigning particular pixels in the block a light colour, and a block can have only one colour at a time. However when we move on to use graphics-based applications we will expect a higher level of resolution and a bigger colour range. The relationship between the display and the information it contains is also important. Most graphics software will use either vector or raster display modes.

In a vector-based application, information about the image is stored in a way which defines individual objects by co-ordinate values which map to pixel positions on the screen. The simplest vector storage unit is a point or node which maps to a single pixel defined by a co-ordinate pair (although in reality it will be made to occupy a block of pixels on the screen in order to be visible). Lines or vectors are defined by linking two or more co-ordinate pairs, and the on-screen lines between them are projected using appropriate functions. Finally areas are defined by linking three or more points into closed structures in the same ways. Quite complex patterns can be built on the screen from relatively little information in store. Only the pixels that form part of the pattern need special treatment. The rest will have no defined value, or perhaps a uniform colour forming a background to the image. The great strength of vector organisation is that it takes an intelligent approach to screen objects. Individual objects or groups can be added, subtracted, or moved without disturbing the rest. The individuality of these objects is emphasised by the common use of the term 'entity' to describe these objects. They can be assigned visible 'attributes' like colour or abstract attributes in the form of data sets. With appropriate programming, vector objects can be made to move and rotate on the screen. Improvement in computing power and memory means that they can be given realistic surface shading, a great advance on the wire-frame models typical even a few years ago.

Raster displays make direct use of the raster scan organisation described earlier. In other words, a typical raster application builds the screen display from information about individual pixels. Although storage might be optimised by providing a group identity for blocks of pixels with identical values, it is easy to appreciate that a high-resolution 256-colour image will need a lot of memory. The bitmap images of raster graphics have other limitations. Because the image is defined in terms of pixel identities, there is usually no way of identifying bigger structures. Individual sections cannot be taken out or manipulated seamlessly in the way that vector objects can, and they cannot be animated in the same way. However the unselective nature of raster organisation, which simply accepts and records the attributes of the pixels that form the whole picture, means that much less prior processing is needed. In this it is quite different from a vector display, which must have structured objects as input. Generally speaking, then, raster images make larger storage demands than comparable vector maps, but have the advantage of being more easily manipulated when loaded. The basic input device of raster graphics is the scanner, which simply records a document or photograph in bitmap terms. Vector graphics depends on the digitiser, a device which allows the user to track a cursor over the base document, selectively recording only those points that will be used to build vector objects.

The two forms tend to have strengths in different contexts. Raster organisation is most useful where texture and detail matters. The most obvious and important use of raster imagery in geography is in remote sensing, where real-world images are put on the screen and then analysed in terms of tonal and textural characteristics. In general terms the patterns of abstract data often used by human geographers are more likely to be suited to vector-based software. The two forms are not mutually exclusive, and it is not difficult to find hybrid applications, where raster imagery is used as a backdrop for vector patterns, or vector patterns are used to superimpose abstract patterns like national boundaries on raster images. As these examples suggest, however, one of the two forms will normally be the dominant mode. Images can be converted. It only takes a few seconds using appropriate software to store a vector map in bitmap terms, and special software also exists which can be used to 'read' raster images and selectively pick out the patterns of pixels which will be used to build vector objects. A confusing range of protocols and file types has been developed for the storage of graphics images. Bitmap files are always stored only in machine-readable form, but some vector formats use text-based files which experienced users can edit without even having the image on the screen.

Computer atlases

During the last few years a sizeable library of computerised maps has been built up, often in the form of atlases. Because these are usually greedy for storage space, they have usually been rather large and needed special handling. In the earlier years they were typically downloaded from tape to mainframes and workstations rather than personal computers, but the development of CD-ROM storage has revolutionised access to imagery like this and put a new emphasis on the PC. These maps are designed to cope with the different needs and technical expertise of distinct user groups, and not surprisingly they vary a great deal in format and appearance. What they generally have in common is a limit to the amount of modification the user can carry out. In some cases they are purely read-only. In most cases the user is allowed to download data or select elements of the image, but does not have the ability to make substantial or permanent changes. A few examples will give some idea of the range.

The *Digital Chart of the World* was originated by the United States Defense Mapping Agency (DMA), and first published on a set of four CD-ROM disks in 1992 with special viewing software. It lets you build your own topographic maps with user-selected features on any practicable scale for any part of the world. These can then be saved to disk or printed, leaving the original data undisturbed. The *Chart* is vector-based, and works in a way which illustrates both the advantages and problems of the vector mode. One of the advantages can be seen without much difficulty. There is an impressively large choice of data types, consisting of about 140 categories grouped into seventeen sets called coverages. The user can build up a map which includes only those data features appropriate to its intended purpose, adding and eliminating categories on the basis of inspection. However one of the problems of using a vector mode will become obvious as soon as the application is opened. Vector construction is a slow process, and the designers of the chart have tried to overcome this difficulty by leaving all the building to the user. Having chosen the part of the globe which the completed map will cover, the user is then confronted by a blank screen, and must use pull-down menus to select elements to build the image. Figure 9:1 shows a map built in this way.

Even on a fast machine, the building process is painfully slow, and it must be repeated every time data types are changed, or if the scale or boundaries of the map have to be changed. With experience one learns to use a skeleton set of categories to define the map's scale and frame, then add the others in groups to minimise redrawing time. There are other problems. Individual elements cannot be selected from types of data. Taking place names as an example, it is possible to change the typeface, size, and

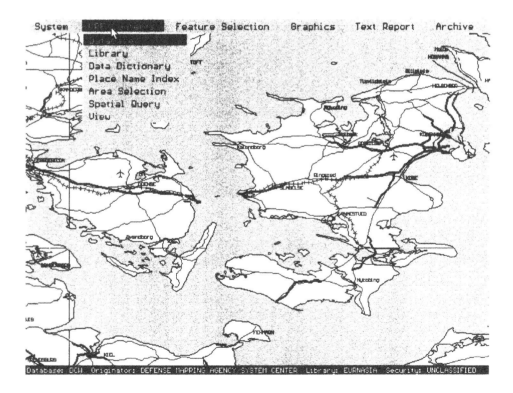

Figure 9:1 *A map built using the* Digital Chart of the World

colour of names globally, but not to move individual names,
edit them out, or give them special emphasis. There is one overriding
difficulty, however, which relates to the issue of generalisation discussed
in Chapter 7. The *Chart* display is designed for a base scale of
1:1,000,000 and does not generalise in response to change of scale. No
matter what scale is chosen, the size, appearance, and quantity of data in a
category remain constant. As a result of this the user's map will tend to
become either cluttered or empty as it deviates from the base scale,
depending on the direction of change.

More recently a number of raster-based atlases have appeared. We can
appreciate the features of these by looking at a representative example,
DeLorme's *Global Explorer*, a Windows-based application which first
appeared in 1993. *Global Explorer*, too, offers a world topographic map,
with a seamless transition in scale from global downwards. In this case
the largest possible scale is 1:200,000, except in selected urban areas
where scales as large as 1:10,000 are possible. One major difference
obvious in use is that the image undergoes a series of generalisation
transitions as scale is changed. These are mainly successful, although
information is rather sparse in the street maps. In this sense *Global
Explorer* really consists of a family of maps designed to be appropriate to
different scales. It offers fewer data categories than are possible on the
Digital Chart, but in compensation offers a series of databases, some of
which can be activated by clicking 'hot-spots' on the screen map, and a
set of search procedures enabling the user to find and move to specified
points. Being raster-based, changes of scale and location are performed
very rapidly. Figure 9:2 shows a screen image from a session with Global
Explorer.

Global Explorer represents a family of applications which have
successfully improved the service offered by the traditional atlas, by
providing scale flexibility and a lot of additional information. As such, its
purpose is quite different from that of the *Digital Chart*. As one would
expect with a raster-based application, the user has no control over the
selection of data types, and images can only be exported as bitmaps using
the routine cut-and-paste facilities of Windows. Raster mapping of this
type now supports commercial applications at a variety of scales. These
range from straightforward scanned versions of existing national atlases
to interactive packages designed with customised output in mind, like
Maps in Minutes for the Macintosh. Street atlases feature prominently.
They are either scanned in from existing paper publications or generated
direct from computer use, and like global atlases they use clickable screen
'hotspots' and database searches to add a layer of information that the
paper form could not provide.

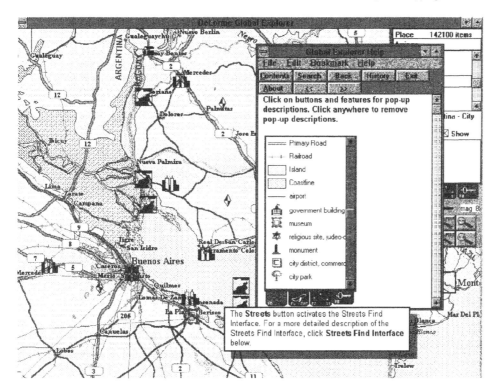

Figure 9:2 *A map built using* Global Explorer

CD-based atlases are also being used to store and present thematic information. Some have an emphasis on environmental issues, and there are also several presenting census and other statistical data. There are at least three CD-ROM presentations of data from the 1991 Census of Great Britain. The possibility that in future large data sets like censuses will routinely be made available in packages capable of immediate spatial analysis takes us into the field of Geographical Information Systems surveyed in Chapter 10.

The Ordnance Survey of Great Britain uses both vector and raster graphics as appropriate. The major digital mapping programme described in Chapter 7 was based on digitisation and vector storage. The large-scale plans on which the programme was founded have traditionally emphasised functional linear patterns rather than attractive shading, and this made vector definition the sensible choice. In any case many of the agency's professional users, such as planners and civil engineers, were used to these large-scale plans and did not find difficulty in coming to terms with their digital equivalents. The digital data are made available in a series of formats which allow the maps to be loaded and modified (subject to copyright) by design software and vector-based Geographical Information Systems. At all scales the professional user has total control over the use and modification of the vector patterns, and on some of the smaller scales users are able to order customised forms with user-selected boundaries and data types chosen from a menu. Because the Ordnance Survey has already carried out the scale generalisation process, users of its digital maps are unlikely to suffer the problems of slow regeneration of images encountered with applications like the *Digital Chart*. It may be significant that this process of vector-based digital conversion was almost complete before the Ordnance Survey became involved in commercial production of raster versions of its maps at popular general-purpose scales like 1:50,000. The Ordnance Survey also launched its own CD-ROM *Interactive Atlas of Great Britain* in the non-specialist market at the end of 1996.

Do-it-yourself computer maps

Most students in geographical fields will sooner or later want to create maps on the computer rather than simply use existing patterns. Traditionally this was a process fraught with difficulties, and there is still at the time of writing a real shortage of flexible mapping packages suitable for non-specialist student use. Demonstration packages like the TLTP GeographyCAL 'Map Design' unit are very useful but are not in themselves platforms for flexible map production.

One of the traditional ways of working was to make use of programming techniques with special features. An advantage of vector storage and display from the geographer's point of view is that by definition vector handling requires some form or other of co-ordinate system to record point locations. Any form of vector graphics is thus in one sense geo-graphic, no matter what form is taken by the finished output. However, the student needs special skills to be able to manipulate the techniques. One rather daunting process was the use of the Graphical Kernel System (GKS), a graphics system standardised in the mid-1980s and designed to be 'bound' to programming languages for implementation. Although nominally not tied to one particular language, GKS was associated particularly with Fortran 77. At a later stage general-purpose program-ming languages were given graphics extensions which allowed GKS-like routines to be run. Figure 9:3 is offered as an example of the kind of graphics possible in such a language, Turbo Pascal 6.0. This simple map depicted here is based on manual digitisation of a very stylised map of England and Wales, and is part of a timed sequence or animation showing

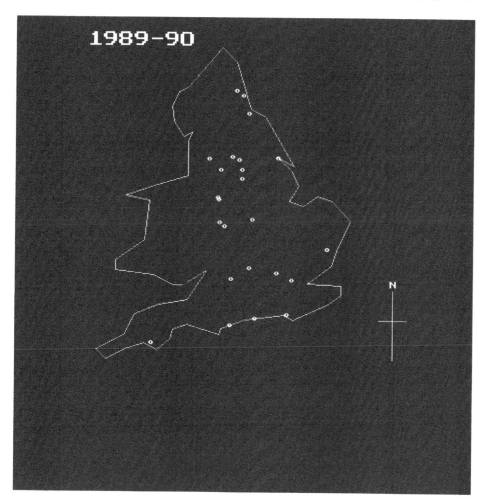

Figure 9:3 *A map produced using Turbo Pascal graphics*
This map is a 'still' from a sequence showing the distribution of the members of different Football League divisions

spatial change in the location of teams in the different divisions of the Football League. Turbo Pascal's graphics supports both vector and raster operations.

Although not very polished, graphics like this can be very useful in handling specific projects. However the overheads of preparing them are forbidding. Learning a programming language to the level of being proficient in rather unfriendly graphics extensions may be wasted effort if more flexible tools are to hand. In the next section we will explore the range of possibilities.

Using mapping software

It hardly needs to be pointed out that every mapping task has its own distinct requirements, and it is very important to be clear about one's needs before starting work. In this section we will examine the main criteria on which a mapping package might be chosen and briefly look at two quite different packages that can be used by students in the geography area without a great deal of training, which should be able to meet most needs.

At the simplest, most students will sooner or later need to produce simple topographic or thematic maps. In the absence of any better software, these can be drawn using a generic pixel-based drawing package like Microsoft Paintbrush, which is part of the Windows package. Quite a reasonable map can be produced with a little care, but working in this way has obvious disadvantages. First, most freehand drawing packages lack co-ordinate reference systems, and this means that the map has to be drawn by eye. This effectively limits the technique to sketch maps and other simple diagrams where accuracy is secondary. Second, pixel-based applications do not tolerate errors or changes of plan. Correcting a minor mistake will mean the erasing of the error and patching of the damage, which can be quite time-consuming. A change of mind about the position, size, or shape of objects may mean even more radical rebuilding. Finally, the completed map is a static item. However good it is, it offers little possibility of use as the base for families of related maps, and it cannot be interrogated except in a simple visual way.

Having established what can*not* be done by a pixel-based drawing package, we are in a better position to establish what we actually need in a mapping package. It should be co-ordinate-based, and ideally have a means of digitising direct to the screen or importing digitised images. It should be object-based to give us the maximum flexibility in modifying the image. If our map is thematic it should also be capable of responsive behaviour. In other words it should be possible to change shading patterns or tints systematically on demand. If it draws its data from a related database it should be possible to redraw it in response to any change in the database itself, and at a more advanced level it would be useful to be able to carry out queries on the database and see the results on the map.

There are drawing packages which meet some of these criteria. Corel Draw, for example, is an object-based Windows package with a co-ordinate base. It thus meets the first two of our criteria, and has the advantage of possessing a very large library of clip-art objects, including well over 100 outline maps. However, it is not capable of accepting direct digitised input, and as a dedicated drawing package it is not equipped with any of the features which would make it responsive. Corel Draw can be very useful, but it does not meet all of our needs.

Computer Aided Design and AutoCAD

One solution is to explore the possibilities of Computer Aided Design or CAD. CAD is a well-established field, with its origins in the needs of architectural and technical drawing, but it lends itself quite well to geographical use. As an example we will look at Autodesk's AutoCAD, one of the most well-established packages in the field, although other CAD packages are broadly speaking able to offer the same tools, and it should not be assumed that any of the features described below are its exclusive property. AutoCAD is a large and expensive product, but Autodesk have recently launched AutoCAD LT for Windows, a simplified version which runs fast in the relatively friendly Windows environment and has a good range of functions. The description that follows applies to both unless indicated.

AutoCAD is essentially interactive. Images can be built up and saved to disk for later use or modification, or sent to a plotter or printer, but the key element is the use of a series of commands within the primary screen area, the so-called Drawing Editor, to build up the elements of the image on the screen. AutoCAD works on the basis that the drawing is a rectangular area, and not all of this rectangle is necessarily visible on the screen at any one time. From the point of view of the cartographer, this has the benefit of allowing work in extreme close-up when detail is important, and rapid zooming out to view the relationship between component parts. Within this framework AutoCAD provides a range of commands for the creation and manipulation of objects. An object can be a straight or freehand[1] line, a polygon of a defined type, a circle, a piece of text, and so on. Objects can be input directly to the screen using a mouse or built up by the use of a command line. Alternatively they can be digitised interactively, or imported in suitable formats from earlier digitising sessions. Even in a complex drawing, all objects have their own line type, defined colour, and other attributes, and they can be re-dimensioned, moved, or rotated quite independent of other objects.

A very important concept in AutoCAD is that of layer. A drawing can have several layers, which are best thought of as overlays. If we were to think in topographic mapping terms, we might have a relief map of an area, a map of land-use features, and a map of settlement and communications, all covering the same area and occupying the same space. In AutoCAD terms these are three layers, which could be viewed simultaneously, or selectively switched on or off. Thematic mapping opens up a wider range of possible layerings, defined entirely by the purpose of the map. The objects on every layer can have the full range of elevations and thicknesses. However every object is a member of a specific layer, and can therefore be selected or removed according to how the image is to be composed. There are also commands enabling the user to obtain on-screen statements of line length or contained areas. Figure 9:4 shows a simple map built up from several layers carrying different information. In this example a retail researcher is plotting the distributions of customers in a street market at intervals during the day. Although Figure 9:4 shows several layers superimposed, individual layers or selected combinations can easily be isolated for study.

AutoCAD works on a Cartesian co-ordinate grid basis. Individual objects have positions defined by co-ordinates within the drawing field. These locations can be defined by typing in co-ordinate pairs on the screen, by using a mouse to identify specific points, or by using a digitising tablet. A visible grid can be displayed on the screen and used for orientation, but

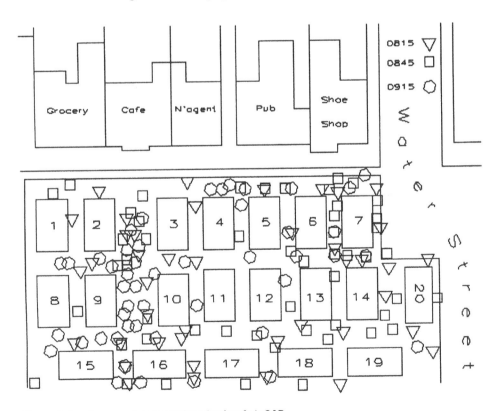

Figure 9:4 *A research map produced using AutoCAD*

this is no more than a guide. Although the grid is measured in units (which can obviously be related to real-life measures), AutoCAD stores the locations of points to several significant decimal places. In the more recent versions there is also a Z dimension, i.e. height.

This means that three-dimensional modelling is a possibility. Every three-dimensional object in an AutoCAD drawing has properties called elevation and thickness. Elevation is the height in units above a base level of zero, and thickness is the actual height in units occupied by the object. Thus when a rectangle is defined in plan as having elevation 10 and thickness 30, if it is viewed in three dimensions it will be seen as a box 30 units deep floating 10 units above the base level. By setting an elevation level, a whole series of objects can be designed at the same height, and if necessary they can be changed individually later. An important concept in relation to the third dimension is viewpoint. The default viewpoint is a simple plan view, but it is also possible to select a specific viewpoint to look at a map or diagram, freely rotated in all three dimensions. Theoretically at least it is possible to create a contour map and treat it as a block diagram by looking at it from oblique angles.[2] A final refinement of direct value to us is hidden line elimination. Any three-dimensional object must have features which are invisible from a particular viewpoint, and hidden line elimination lets the user get a clearer picture of an object's overall structure by a process which identifies and suppresses the line-work making up any hidden parts of the three-dimensional image as seen from a particular viewpoint. With hidden-line elimination it is possible to make modelled objects look rather more like their real-life counterparts.

AutoCAD also provides a useful platform for the handling of Ordnance Survey digital maps. These are available in a number of different formats, one of which is Drawing Exchange Format (DXF), and this form is directly readable by AutoCAD after a short import procedure. Viewed in AutoCAD, Ordnance Survey maps are organised into a large number of individual layers, each with its own colour code and other attributes. When the file is first opened all layers are visible, but the user can turn them off as desired, and add entirely new layers which can be saved as part of a modified copy of the original map. The map image can also of course be changed in scale and shape. Subject to the rules of licensing requirements and copyright, the original digital input can be simplified, annotated, updated, and overlaid with data relating to surveys or research projects. Processed in this way, Ordnance Survey digital maps have a lot of potential for the rapid production of high-quality plans for project and dissertation work. However it must be said that Ordnance Survey digital images are not cheap, whether bought outright or licensed, and this means that for most students the only maps realistically available will be those already licensed by their college.

It is not difficult to learn enough of the basic AutoCAD program to provide diagrams and plans up to a reasonable standard of presentation, and it is ideal for projects in which vector-defined objects are the dominant form. Like all software, it has its limitations. For example, raster operations like shading and texturing are not easy. However, there is also a family of supplementary and allied programs which improve the quality of the image and provide a very large range of additional design tools, although most of these are too specialised to be valuable to us here.

MapInfo

Although superficially similar, MapInfo[3] is a very different application. AutoCAD in its basic form is simply a descriptive package, and the meaning of the images created by it depends on the pattern of objects on the screen. However MapInfo relates maps to patterns of data. A typical MapInfo operation will involve the matching of a previously designed base map to a data file containing information that describes the characteristics of areas, points, and lines on that map surface. The base map consists of vector-defined objects and the data file contains their attributes. Processes of analysis and presentation allow finished maps to be produced from this basic raw material. By linking screen objects to a flexible database in this way, MapInfo takes an important step further. In some respects it is a form of Geographical Information System (see Chapter 10), enabling the user to carry out analysis which may well produce results more interesting than the basic components of the patterns themselves. However MapInfo prefers to describe itself as a *Desktop Mapping* package, enabling the user to visualise data which would otherwise be stored and used only in the context of a non-spatial database. Objects can be defined interactively on the screen or with a digitiser, or imported in an appropriate format. Formats accepted include Drawing Exchange Format (DXF), which we met earlier in the context of AutoCAD. As far as data are concerned it is designed to accept files created in commonly used database management systems, and users can thus put their data into a spatial context without having to go through the expensive process of completely remodelling their files.

As a Windows application MapInfo makes full use of the ability of Windows to support a series of open windows simultaneously, with one at any time being 'live' or active and the others dormant. Among the most significant components of MapInfo are 'files' and 'tables'. In MapInfo's

terms a file is a structure containing information. The basic files used in
MapInfo will contain either raw data or information about the structure
into which these data have to be placed. In practical terms, working
sessions with MapInfo always involve tables rather than simple files. A
table consists of at least two files with a common spatial identity, and the
minimal form of table will have one data file and one structural file.
Opening a table makes available for use a database which has been tied to
a spatial co-ordinate pattern so that it can be viewed as a map. Typically, a
MapInfo session will involve opening a number of files which share the
same co-ordinate system, so that they can be overlaid to form a composite
map. Quite complex tables can be created and their patterns analysed. The
new composite tables can also be saved for later use.

Another important MapInfo component is the workspace. This is the
working area at a particular stage of a MapInfo session. It contains every
MapInfo element in use at the time, so will probably include several
windows of different types in different relationships to each other on the
screen, although they will not necessarily have much relationship to each
other conceptually. When a saved workspace is opened, it will be restored
in exactly the form that was saved. Saving a workspace is therefore a very
important element in progressing with MapInfo, since it eliminates the
need to start from scratch every time a task is undertaken.

Any open table can be viewed in one of three different ways, through the
Mapper, the Browser, and the Grapher. These can be selected from
Mapinfo's Window menu. The Grapher allows the production of a range
of types of non-map graphs based on the database. These are not unlike
the graphs we will have met in Word for Windows and Quattro, covering
much the same range of performance and being produced in much the
same way. For that reason it is not covered in detail, and we will look
more closely at the Mapper and Browser.

As the name suggests, the Mapper allows the file or table to be depicted
and modified as a map. A typical MapInfo operation is to build up a
complex map by adding in a series of different tables with the same co-
ordinate base but different data values. Although they may be overlaid on
the screen, they do not lose their individual identity. Each component
table becomes a distinct layer of information, which can be added,
modified, or removed from the composite table as appropriate. Even when
tables have been merged in this way, MapInfo lets the user open separate
windows in the workspace containing the individual component tables so
that they can be viewed separately.

When a map is open in MapInfo, there will always be at least two layers.
Apart from any map table layers containing vector objects, there will

always be the rather oddly named 'cosmetic layer'. This cosmetic layer is the one which contains labels and usually other text information. MapInfo provides a Layer Control mechanism to add and remove the layers, change their individual specifications, or overlay them on others. Layer control can be used to make layers selectable, meaning that the mouse can be used to pick and highlight their individual objects. The map in a window may at first not be quite the required size, but it is easy to zoom in and out as appropriate.

The Browser opens up the database for inspection and analysis. Every object recognised by MapInfo is assigned an identity number, which can be established by using the so-called information tool. When this is active, clicking an object will display its identity number and database attributes. However the Browser allows a much more systematic appraisal of the whole database. Each map object has a matching database record, each of which contains one or more fields. The data type, size, and other characteristics of each field are normally defined by the user before data are entered. Modification is possible at a later stage but may risk data loss, particularly if fields are truncated. Data can be imported in bulk from other databases, but for small applications it is possible to define fields and assign attributes to them direct from the keyboard.

Most of the value of MapInfo comes from the ability to interrogate tables in ways that display on the screen the results of queries focused on the database records in the Browser, using Mapinfo's analysis tools. A simple example might give some idea of the range of possibilities. Figure 9:5 shows the screen view of a working session involving a survey of recreational activities in Epping Forest. The map itself records as polygons six areas defined for the purpose of the survey. The Browser window summarises the database of scores derived from the field survey. Only part of the database is visible in this reduced box, but the Browser could be given full-screen status if we wished. One area – the Warren – has been selected by selecting the Pick tool in the Tools menu (highlighted) and clicking on the polygon. The chosen polygon is identified by a hatched pattern and at the same time the appropriate database record in the Browser is also highlighted.

We are not confined simply to displaying patterns and matching data attributes. For example it is possible to use shading to produce a map of different values for one of the database variables. The Analyse menu offers a 'Shade by Value' option, and we could use this to produce a map in which the polygons were shaded according to their scores on one of the database fields. When this option is chosen and we open the Shade by Value dialogue box to identify the layer which is to be shaded and the one

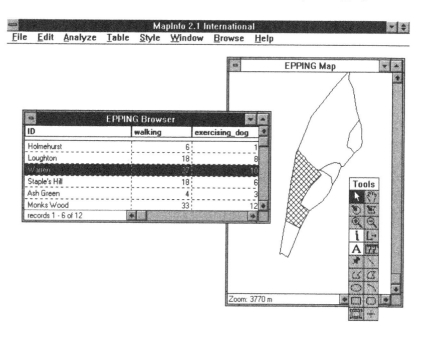

Figure 9:5 *A map built using MapInfo*

from which the values are to come. We can then let MapInfo select a different colour for each value, or carry out the more constructive process of grouping values into our own defined number of categories. When all options have been finalised, the map will be redrawn. If the first image is not satisfactory, we can experiment with different shading tones and textures until we are satisfied with our choice.

We can also carry out a search to pick out cases from the database that meet particular requirements. Suppose that we want to find out whether our survey areas are not equally attractive to people exercising dogs. Here again the analysis tools are used. In this case the Select option is used, and it is possible to build up conditions using field names, logical operators, and target values. At the very simplest, we can ask MapInfo to identify polygons in which dog-walkers exceed a defined number, but we can refine the analysis, perhaps by assessing dog-walking in relation to other recreations or polygon areas. When the analysis is complete, MapInfo will display the map window, with any areas that meet the condition highlighted. The Browser will also be displayed, and here too the target group will have been highlighted. They can also be seen as a discrete selected group if necessary.

As a final example, it is possible to search the map for a particular item and mark it with a symbol. Suppose that we had also recorded a layer of data showing refreshment sites and other attractions that might influence the movements of visitors to the Forest. The analysis tool will allow us to search the database table to identify the name of a chosen point. This operation (which can be repeated for any number of cases) overlays a marker symbol on the symbol that already exists on the map. This marker can now be used as the centre of an area search. MapInfo has what it calls a Radius Search Select tool. When this is activated, the mouse can be used to drag open a circle based on the point marked by the symbol. As soon as the mouse button is released at a suitable radius, all other attractions within the circle will be highlighted. There is also a Polygon Search Select tool which lets us highlight all the sites within a particular polygon. These simple radius search tools thus let us obtain extra information. We now can define sets of attractions defined entirely in relation to the position of a chosen point. When a selection like this has been made, spatial analysis can be done on the selected group alone, and a new database table is produced containing only the attributes of the highlighted points.

These examples are extremely simple, but they do demonstrate the way in which MapInfo can be used. It can be seen as a cartographic tool in the narrow sense, allowing the user to view and select shading and symbol patterns for a map which can then be plotted or printed. However there are also important analysis elements, and here the power of MapInfo lies in the ability to interrogate the database and map the pattern of selected attributes. The new patterns themselves can of course be shaded or symbolised as necessary. It does not take a lot of imagination to realise that a series of analyses can be performed on successive selections to analyse relationships that would be far from obvious when the original data were viewed.

An exciting application now available with the most recent version of MapInfo (Issue 4.1) is Vertical Mapper, an optional extra which cannot be used outside the context of MapInfo. Vertical Mapper lets the user create contour or gradient-shaded maps from data and overlay them on other tables. It can be used to produce output very quickly and effectively by selecting sets of data from the Browser and processing them as appropriate. Naturally enough it will produce much better results for users who take the time to learn about the full range of its facilities. An obvious application of Vertical Mapper in our Epping case would be the production of maps showing contour or density-shaded maps of different activities based on data recorded at specific points within the polygon pattern.

Summary

This survey of the field of geographical applications has concentrated on mapping techniques. Major improvements have taken place in this field in the last few years, and students now have access to a powerful battery of mapping tools. The allied developments of hypertext (see Chapter 11) and mass storage forms like CD-ROM promise to produce a lot of very exciting material in the next few years. On the other hand a cheap, flexible, and easy form of computer mapping for basic student work is still not available.

Not all geographical software relates to mapping, of course. Modelling is an important field, and this chapter would not be complete if it did not at least mention the existence of general-purpose modelling packages like Stella II for Windows by High Performance Systems. However the field of modelling is too big and complex to be handled at this stage.

Discussion questions and tasks

1 Most British universities with Geography departments will be able to give you access to the GeographyCAL Map Design module. Explore it.

2 What other mapping software is available to you locally? Use the text to assess the possibilities it offers.

3 Pick any existing map. Select the software you think you might need to make a digital version of it (simplified as appropriate), and plan the sequence of operations you will need to complete the job. Then do it!

4 Think of the ways in which MapInfo lets you link databases to maps. Then go and read Chapter 10.

Further reading

See also

Computer techniques, Chapter 8
Geographical Information Systems, Chapter 10
The Internet, Chapter 11

General further reading

This is a field in which reading tends to be either very generalised, or very specialised in the form of user manuals. One of the better general guides to the topic, although now rather dated, is:

Mather, P. (1991) *Computer Applications in Geography*, Chichester: Wiley.

The *Cartographic Journal*, published twice yearly by the British Cartographic Society, carries a lot of information about computer mapping and is certainly worth a look.

Notes

1 AutoCAD LT does not support freehand linework.
2 However there are problems. The hidden line command that is used to produce clear three-dimensional views does not work with lines of zero thickness.
3 This description is based on MapInfo 2.0. Like all software, later versions will differ in detail.

⓾ Geographical Information Systems

The development of Geographical Information Systems has made the world at large realise that there is more to geography than capes and bays. This chapter covers:

- The nature of GIS
- Types and structures of GIS
- The phases of a GIS project
- Some thoughts for the future

Geographical Information Systems or GIS[1] have a history that goes back in one form or other to the 1950s, but their expansion on a large scale was a phenomenon of the later 1980s and early 1990s. GIS developed the power to answer important geographical questions, questions which are not asked exclusively by geographers, and GIS came into use across a wide range of different disciplines. By the early 1990s it was no longer a technology that supported existing fields, but had become a discipline in its own right. It now supports a large literature including monumental works like the two-volume *Geographical Information Systems* edited by Maguire *et al.* (1991) and its own journals like the UK-based *Mapping Awareness*. Some academic geographers who did not know how GIS were organised and implemented seemed to regard them as almost magically powerful tools. The rise of the Internet may have provided a new focus for this worship of technology, but GIS have quietly gone on entrenching themselves more firmly as the 1990s have progressed.

The expansion of GIS has not been spatially uniform. The biggest early developer of GIS techniques was the USA. This is not simply because of the country's domination of development in the computing area, but relates to the organisation of American space. The USA has an exceptionally large amount of public land which requires recording and administration, and this is carried out by a wide range of federal or state agencies. Collectively they provided a booming market for suitable software. The physical area of the USA is also huge and has clearly defined regional contrasts which make it difficult for land managers to ignore the spatial differences between or even within the areas they control. In contrast, the UK has a small public land proportion, and one which has actually been shrinking during the last few years as the boundaries of state control have been narrowed. The UK has a much more centralised apparatus, and is a small country with less obvious regional contrasts. Both of these factors tend to favour the use of traditional non-spatial management software.

However GIS is now genuinely global in scale, in two senses of the word. Some GIS have a literally global scope and monitor patterns world-wide. The application of GIS on a local basis has also become more or less universal. Technical improvement means that GIS have migrated from mainframe machines down to workstations, and even a reasonably powerful PC can now be an adequate host. The combination of perceived need and technical flexibility means that there can be no more than a handful of countries world-wide which do not possess at least one locally based GIS. In most parts of the world GIS penetration has tended to start with one of a handful of user types such as national cartographic agencies, land registers, or (less commonly) environmental management agencies.

For a number of reasons this review will be brief. Although it is important to understand the principles and logic of GIS, the details of their operations are important only if they are to be used practically. Demonstrations of GIS applications with prepared data sets are sometimes used in the earlier stages of degrees in human geography, but the size and normally rather unfriendly nature of full-scale GIS means that teaching does not often go beyond this level except in specialised options. GIS are not tools to be picked up casually for instant use. The heavy overheads of training and data input involved in making constructive use of GIS also rules them out for all but a minority of student research projects and dissertations.

The nature of GIS

The development of computer handling has meant that information has become an important and expensive commodity in its own right, and this has promoted the development of a formal Information Science handling information systems. Nobody serious disputes that GIS belong legitimately in this field of information systems. As such, we can expect that all GIS will share the general characteristics of information systems, which can be summarised as dealing with data input, handling and output, management, updating and upgrading facilities, and training for the user community. Without labouring the point, there is equally little argument about the way in which GIS are distinct. They are special members of the family of information systems because their information has a geographical component. We can assume for the moment that 'geographical' means that the data relate to points or areas within a spatial referencing system. This might be the purely relative ordering given by an arbitrary grid, but it is much more likely that we will use real-earth co-ordinates such as the latitude/longitude system or the UK national grid. Perhaps in one sense this word 'geographical' is the wrong one. 'Spatial' might be better, not least because as our interests develop we will start realistically applying GIS techniques to areas beyond the boundaries of our own planet, where the term geographical is no longer appropriate. However it is too late to change that. The term GIS is too firmly rooted.

There is more to a GIS than simply spatial referencing. Every spatial organisation has latent meaning. As we saw when we looked at the paper maps which are the ancestors of GIS, they were always produced with an aim in view. That aim was to display data spatially to provide a perspective that could not otherwise be obtained. With GIS too, spatial referencing is a means to the larger end of purposeful data analysis.

How do we define GIS? A well-established definition in a British context was produced by the British Government's Committee of Enquiry chaired by Lord Chorley (the Chorley Committee) in its report on *Handling Geographic Information* published in 1987. It describes a GIS as 'a means of capturing, storing, checking, integrating, mapping, analysing, and displaying data spatially referenced to the earth's surface' (DoE 1987: 312). This definition has the merit of stressing the structured sequence of processes that contribute to GIS analysis, but it does not provide a very clear idea of the reasons for carrying out this kind of analysis. A simple definition offered by the Environmental Systems Research Institute (ESRI), which produces a major GIS called Arc/Info, is that a GIS is 'a toolkit for the manipulation and interrogation of geographical entities and their associated attributes'.[2] Here the emphasis is on the central importance of analysis rather than individual components.

The difference between these two definitions suggests that there is room for debate, and we do not have space to multiply definitions unnecessarily. After surveying a large number of attempts, Maguire found that the one common feature of the definitions he examined was that they regarded Geographical Information Systems as systems for dealing with geographical information! However Maguire also made a very important point which many definitions ignore or at best leave implicit. In GIS reality is examined in terms of the relationship between geographical and attribute elements (Maguire 1991: 10–11). The primary characteristic of geographical elements is of course that they can be defined in spatial terms. These 'geographical' or 'locational' elements are also sometimes called 'entities' or 'objects' (see Chapter 9).[3] Attributes (which we also met in Chapter 9) are data which may not have directly geographical characteristics of their own, but which can be mapped on to the entities. Other terms such as 'statistical' and 'non-locational' are sometimes used in place of 'attribute', but in this text 'entity' and 'attribute' will be used throughout to refer to the spatially and non-spatially defined elements respectively.

In terms of human geography, the spatial organisation of census information provides a good example of the entity–attribute relationship. In Great Britain the basic element of data gathering and analysis is the Enumeration District or ED. There are a great many of these – over 100,000 in England and Wales in 1991 containing an average of about 180 households each (Dale and Marsh 1993: 226). EDs are entities recorded simply in spatial terms, and each one is in itself no more an empty shell defined by the co-ordinates of its boundaries. However each ED is also the host to a large number of attributes based on the twenty or so questions asked about each household member in the Census. It is thus a spatial vehicle for analysis of distributions revealed in the Census. The relationship is symbiotic. Collection of the attribute data provides a purpose for the creation of the spatial framework, and the spatial element adds greatly to the power of analysis.

All forms of multivariate analysis have the power to use the relationships between individual variables to reveal relationships that were latent in the original data set. One of the most exciting things about GIS is that it lets us do this in spatial terms. A GIS analysis can legitimately generate entities that did not exist in the original data set. In the Census context, for example, an analysis of the relationship between (say) ethnic identity and household amenities might identify distinct groups of EDs that can be aggregated to form new entities, and these in turn can be mapped or used in further analysis. As we shall see later there are also GIS tools that can create entities with entirely new boundaries.

Types and structures of GIS

In every GIS the ways in which data are captured, analysed, and displayed on the screen or in hard copy will primarily be based on vector or raster organisation. These modes are not mutually exclusive, and most GIS have the ability to make use of supplementary data organised in the unfavoured mode. In Chapter 9 we looked at the difference between raster and vector data organisation, and came to the conclusion that the software used by human geographers tended to be vector-rather than raster-based. This is true of the GIS field as well.

Because raster data organisation is based on the handling of data contained in individual pixels it is best suited to continuously varied distributions in which the key data values are embedded in an image where every pixel carries an individual value. A typical raster GIS application is one where large numbers of points contain discrete data values rather than those where large numbers of values are concentrated on a limited number of locations. For this reason it is most typical of analyses where photographs or satellite images are the primary sources of data for processing. The assembly of different types of data for comparison and analysis will normally be done by processing different images with the same spatial reference points. This dependence on tangible imagery limits its value in most fields in human geography. Pixel-based storage has advantages for some kinds of analysis, particularly types of simulation used in physical systems. This makes it useful in land assessment, agriculture and some environmental fields, but it is not really appropriate for a great many human analyses. A raster GIS like Idrisi, which has a large repertoire of specialised image analysis techniques, would be ideal in physical cases but it would not be a good choice for most other purposes.

Human geographers are much more likely to use data in abstract forms, and they will often be available for points or bounded areas rather than continuous surfaces. The case of the Census which we looked at briefly earlier is a good example. The Small Area Statistics of the Census for Great Britain are based on Enumeration Districts and provide 86 tables for each ED with an overall total of about 9,000 statistical counts for each one (Dale and Marsh 1993: 205). Each ED therefore potentially has a sizeable database of attributes tied exclusively to it, although directly comparable with those of other EDs. All of these data are derived from the original Census Returns and none could be derived from photographic imagery or other physical recording methods. The choice is therefore likely to fall on a vector-based GIS. In this case data will be input in two distinct phases which require quite different input modes. One stage is the construction of the pattern of entities, which needs the definition of the points that delimit each entity. The other is the assembly of their attributes. This will probably be a conventional tabular database process. Because a whole series of attributes can be tied to an entity simply by ensuring that the data set and entity carry the same identity number, there is no need to provide spatial references in the database. Attributes can be added and deleted systematically without any need to do any further spatial marking. A well-established vector GIS widely used in universities is Arc/Info, and this is the example we will look at later in this chapter.

Vector-based GIS typically have structures which reflect this two-fold focus. At the heart of the structure is the relationship between geographical or entity data on one hand and statistical or attribute information on the other. In the case of Arc/Info the name itself reflects the duality. Arc is the module that handles the spatial mapping, and Info is the database module. The classic GIS pattern, as demonstrated by a well-established application like Arc/Info, is for there to be a specialised database management system. At the heart of the structure is the database. Like any other well-designed database it is capable of receiving primary input and updates, and it has its own database management system or DBMS. The special feature here is the linkage established between graphical data used to construct the map and the attribute data describing the information items.

The phases of a GIS project

Like all types of research, GIS work involves a process which carries through from the first conceptualisation to final presentation. Table 10:1 shows how the sequence runs.[4] Not surprisingly, the underlying structure is like the one summarised for research projects in general in Chapter 2.

The design of the database is vital. At this stage the purpose of the project, the types of data to be used and their relationship to each other, and the logistics of data generation all have be taken into account. Decisions taken (or omitted!) at this stage will largely dictate the way in which the rest of the project develops.

The input of data is also very important. Data capture on a scale large enough to be meaningful can be a slow enough process to be a major bottleneck in GIS research, and it can also be very expensive in terms of time and money (Maguire and Dangermond 1991: 324). This is a major reason why serious independent work using a GIS is unlikely to be practicable for non-specialised human geography students. The modes of data capture used by GIS normally use one or other co-ordinate system to establish the spatial position of entities. Typically this will be a standard xy co-ordinate system. A vector-based GIS like Arc/Info recognises three forms of entity data. Arc/Info gives these the collective name of

Table 10:1 *The phases of a GIS project*

Building the database

      Designing the database

      Inputting spatial data

      Editing and creating topology

      Inputting attribute data

      Managing and manipulating data

Analysing the data

      Exploring existing relationships

      Defining new data relationships

Presenting results of the analysis

      On-screen maps and reports

'features'. These are the familiar categories of points, lines, and areas, defined by increasing numbers of co-ordinate pairs as we saw in Chapter 9 and increasing in dimensionality from zero through one to two. In the terminology used by Arc/Info lines are known as arcs and areas are called polygons.

The actual capture of data can be done in one of three main ways. The most familiar is by keyboard input, which involves no more than direct input of the co-ordinate pairs required. Data of this type can be captured indirectly, of course, by reading files in which the co-ordinates were set up in advance. Keyboard input of co-ordinates is painfully slow and error-prone, and digitising has obvious advantages for large amounts of data input. In this case too it is possible to get pre-digitised files such as the ones available from the Ordnance Survey in Great Britain and the United States Geological Survey (USGS) in the USA. A third technique is scanning, in which an electronic scanner is used to obtain intensity values for every pixel in a defined area. For use with vector structures this requires an intermediate stage in which the input is processed using an algorithm which uses chosen values in the intensity gradient to define the entities.

Data have to be put into a meaningful form. The working procedure is typically to structure the input of spatial data so that data relating to one particular type of entity is stored in a defined set which Arc/Info calls a 'coverage'. A GIS model of health care provision might therefore start with three separate coverages showing urban population areas (polygons), access roads (arcs) and the locations of hospitals (points). When it is captured every entity will be recorded in the database as a record within the coverage to which it belongs. This record will contain a set of standard attributes. In the case of a polygon, for example, these will include its perimeter length and area and an ID code. A single coverage will normally contain only one data type, but there is no limit to the number of coverages with the same or different data type which can coexist within a set of boundaries.

The editing and creation of topology takes place as the pattern of coverages is built up. The correct use of spatial data in a GIS depends on correct topology, i.e. the spatial relationships between features or entities. Digitised data will normally have to be verified and cleaned of bugs. There may be arcs that do not connect, polygons that do not close properly, and other difficulties. Arc/Info and other GIS have special commands which allow their users to carry out this task. Apart from troubleshooting, there are three positive aspects to topology in this context. The GIS will store cases of adjacency, where entities can be treated as adjoining each other. Where there is connectivity, i.e. where lines meet or intersect, it will store the locations of the nodes where they meet and other information such as distance and direction. It will also store cases of containment, where a feature of any kind is totally within a polygon.

The input of attribute data is a separate process and follows a procedure very similar to that used with relational databases. This is not accidental. For obvious reasons GIS designers have tried to keep attribute input within familiar territory and make it possible for existing databases to be imported. The special feature of any Arc/Info attribute database is that it will contain a key attribute or attribute set from the spatial data record. In this way the new attributes can be matched firmly to the existing spatial tables. Figure 10:1 shows a simple coverage consisting of a set of polygons, and the attribute table containing a series of data items from each one.

The last preparatory stage is the management and manipulation of the data, At this point final coverages will be produced by integrating different data sets. The coverages will also have to be positioned correctly relative to each other, and this is ensured by the use of common ground control points, which Arc/Info calls 'tics'. A good deal of the topological groundwork involved in defining the relationships between coverages also takes place at this stage, when the final coverages are defined.

So far the jobs involved in preparing the database, important and time-consuming as they are, have been no more than preparation for the real essence of the task, which is analysis. It should be obvious by now what is going on; we are preparing what we can envisage as a fully related set of overlays over a simulated ground surface, which can be compared in ways that we see as appropriate to produce the maps which will be our finished products.

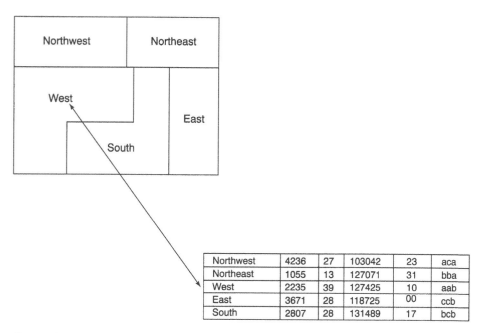

Northwest	4236	27	103042	23	aca
Northeast	1055	13	127071	31	bba
West	2235	39	127425	10	aab
East	3671	28	118725	00	ccb
South	2807	28	131489	17	bcb

Figure 10:1 *A coverage and its associated attributes*

The analysis of data by GIS takes several forms. Some forms of tabular analysis could be carried out on the attribute database without the spatial context of the GIS. The more exciting forms of analysis are the ones which make use of the special features of GIS. These can be grouped into two sets, overlay analysis and proximity analysis (or buffering).

'Polygon overlay' has a number of forms which should be familiar to readers who have studied set theory. If we visualise two coverages as separate overlay maps of the same area, overlay techniques let us create new coverages combining elements of the existing ones. The simplest relationship is the operation identified in Arc/Info terminology as 'identity'. In an identity relationship a polygon coverage is overlaid on another coverage which might be any one of the three types; all the input features are kept. Thus if we overlay one polygon coverage on another, we create a new coverage with as many new polygons as are defined by the overlap of the two coverages, within the boundaries of the overlaying one. Each polygon will have the sum of the attributes belonging to the shared parts of its two parents. 'Union' and 'intersect' overlay operations can be carried out. In the union of two polygon coverages a new coverage is created which contains all the areas within either of the parents. In intersect the new coverage contains only those features (of any of the three types) which are within the boundaries of both coverages. Figure 10:2 shows both vertical and oblique views of a

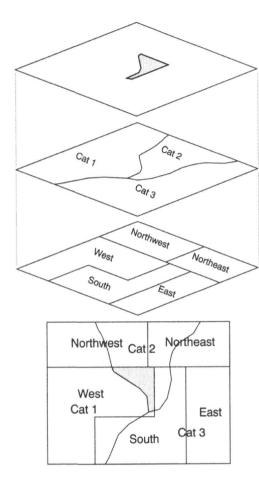

Figure 10:2 *An intersect operation on two coverages*

set of coverages. The shaded area in the upper layer is the intersect of West and Cat2.

Buffering allows us to define a zone of a chosen width around a feature. It can be applied to any of the three major forms. Once again it allows us to create a new coverage, but this time one containing only territory which meets a distance criterion. The significance of the buffer should not need too much explanation, and the examples which follow should be self-evident. We could for example have buffers of 5 km on either side of rivers with irrigation potential, or 12 km around the coastline of the UK to show the status of the fishery zones, or 25 km around the sites of major international airports to identify areas exposed to noise pollution. Figure 10:3 shows a simple buffering operation. The left-hand side shows a pattern of motorways, but the right hand side has been buffered, and shows a 10 km wide zone around each road.

In advanced analysis these operations will be used cumulatively. We might first of all use buffering to identify polygons which are within a certain distance of particular features, and create a coverage which keeps only the buffered areas (or if appropriate discards them and keeps everything else). Then we can take this new coverage and overlay it with a different one of our choice to create yet another. Finally we can perform tabular analysis on this last coverage to create attributes which provide us with the answer to our initial query. All the time we will be using the locational information implicit in the database to create visible maps and tabular reports on the screen and ultimately in output. In that way we complete the process started with our initial concept. Figure 10:4 shows graphically the result of carrying out an analysis of health data.

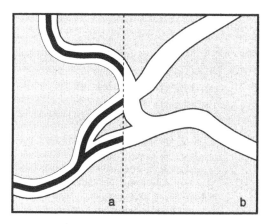

Figure 10:3 *A buffering operation on a road network*

What kinds of analysis are actually done with GIS? A glance through the recent literature emphasises use by local and national government agencies and commercial users. GIS is not only applied to fields like land registration. It has important applications in transport handling, ranging from airport management through road traffic control to accident recording. It is also widely used to map and co-ordinate the web of sub-surface power and water lines controlled by the utility companies. Social applications include local authority housing department inventories and allocation systems, and the management of health provision. Census data play an important part in these fields. In education, the Open University now runs a GIS application designed to target demand for distance teaching and learning.

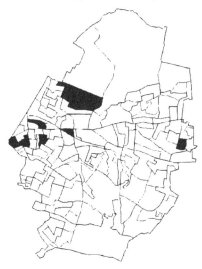

Commercial applications are numerous. Business users were quick to realise the value of GIS both to define under-exploited spatial markets for their products and to help them locate their retail outlets. GIS is now used to underpin the marketing of a wide range of products from beer and burgers to toys.

Figure 10:4 *An example of an analysis using Arc/Info*
This diagram show the results of an exploration of a search for relationships between living conditions and mortality, using the Leicester data from Hickin et al. (1991)

Some thoughts for the future

There is no doubt about the power of GIS and its early expansion was very fast. However, at the start of the 1990s the first flush of enthusiasm began to wane. There were several reasons for this. To some extent GIS was 'oversold' and users were led to expect too much. As a result they sometimes found the results disappointing, and the process also often turned out to be more expensive and slower than anticipated. The biggest problems were not with hardware and software, and GIS made a successful transition to micro-computer platforms. Most of the cost of this operation (as a rule of thumb about 70 per cent) turned out to be in the field of data collection and input. Traditionally input has been by digitising from paper maps, which is a slow and expensive affair with high costs in terms of checking and verification. The development of high-quality scanners and vectorising software to convert the scanned images has been an important innovation. There are other important costs in terms of dedicated machines and accommodation, systematic data update, and user training.

The legal position of GIS work had to be defined. In the United States there have been a number of embarrassing legal cases where planning agencies have been sued because their decisions were made on the basis of negligently assembled GIS data sets. One of the first such cases involved a property-owner in the New York area who was refused authority to build because GIS data had been wrongly encoded so that the site was incorrectly identified as being so close to sea level that permission was automatically refused on the basis of flood risk.

There is also the issue of data quality. Some environmental inventory exercises, including some carried out by agencies in the Third World countries, have been regarded sceptically because, although the GIS set-up itself was faultless, there are grounds to believe that the data collection exercise was badly flawed. Analysis carried out on the basis of defective data is dangerous, and doubly so if the customer is convinced that the computer-generated output must by its nature be fully authoritative.

Finally, there has been debate within the geographic community about the academic value of GIS. Some critics think that it is 'stuck in a rut', concerned with description and the analysis of two-dimensional static patterns rather than modelling of processes and three-dimensional patterns. Supporters concede that there are some grounds for this criticism but regard the situation as the understandable failing of the pioneer phase. The challenge of broadening its scope is still to be addressed (Davies 1995).

There are signs now that this period of uncertainty is over, with an annual take-up in commercial environments increasing world-wide by about 60 per cent in recent years. However GIS is widely seen as the domain of large commercial and governmental users. There is still a lack of GIS small and easy enough to be used by small businesses and (more important in our context!) students without special training. Convergence might provide an answer to this problem. Major software houses with an eye open for market penetration are working currently on ways of building GIS elements into database and business software suites. If this is successful, we may see a radical change in the next few years (McGinn 1996).

Summary

Despite the problems discussed earlier, GIS certainly has a future, and an important one, not least because it creates spatial awareness outside the traditional geographical sciences. At the moment, however, everyone is still coming to terms with the values and limitations of this rapidly growing discipline. Even if undergraduate human geographers may not be able to do much serious work with GIS techniques it is important to be aware of their potential at more advanced levels of research.

Discussion questions and tasks

1 Think about the ways in which GIS analysis can create data sets greater than the sums of their parts.

2 Browse through the journals. Assess for yourself the range of areas in which GIS are being applied.

3 How many of the cases you find are parts of programmes of research in human geography?

Further reading

See also

Data sources, Chapter 3
Computer techniques, Chapter 8
Computer mapping, Chapter 9

General further reading

Maguire, D., Goodchild, M.F., and Rhind, D.W. (1991) *Geographical Information Systems*, Harlow: Longman, 2 volumes.

Do not try to read all of it! All big edited volumes like this tend to be uneven, but the early chapters are helpful and browsing through the rest will give you some idea of the range of the field.

Mapping Awareness and *GIS Europe*

These sister journals are published ten times per year, and are useful browsing sources. Both cover the whole spectrum of GIS applications, concentrating on the British Isles and continental Europe respectively.

Notes

1   In American usage Geographic is used in place of Geographical. This text avoids the trans-Atlantic dilemma by following the convention of the professional literature and using the contracted form GIS as the norm. According to context GIS can be a singular or plural noun, an adjective, and can refer to specific applications or the discipline as a whole.

2   This is the definition used in ESRI's unpublished training documentation.

3   Technically an object is the digital representation of a spatial entity, which could of course be described in non-digital ways as well.

4   The table is based on a figure in an Arc/Info text (Hickin *et al.* 1991), but it would be appropriate for any vector-based GIS.

⑪ Using the Internet

One of the most impressive features of the last few years has been the growth of the Internet. This chapter looks at the growth and covers:

- **The development of the Internet**
- **Getting started**
- **The Web**
- **Gophers and gopherspace**
- **Mail**
- **Telnet, anonymous ftp, and other file transfer techniques**
- **Talk and interaction**
- **Referencing the Internet**

The Internet is one of the most exciting and significant developments in computing of the recent past. Although it has been developed from communications systems that are quite old by computing standards, the revolutionary feature of the Internet is the speed with which it has opened up remote communication within a much larger user group than ever before without requiring much specialised expertise. There is a paradox here. Underlying the Internet is a very highly disciplined system of communication, but in human terms it is not unfair to describe the Internet as a loose anarchic network of users with no overall control or purpose. Indeed for some users this is its charm and value. The evolution of the Internet in the first half of the 1990s was very striking and it is still continuing at an impressive speed. Inevitably, extravagant claims have been made about the impact it is likely to have. It has even been argued that it will change human behaviour in a radical way, removing the discipline of space and creating the paradox of a spaceless geography.

At the moment we cannot be sure where it is going. The Internet is host to academic and corporate data, commercial services, enthusiasms and hobby interests, and a mass of individuals who look to it for fun, companionship, and perhaps a sense of identity. Whether it will be able to expand without losing this richness is not yet clear.

In this chapter our first priority is to look at the Internet, or the Net as it is familiarly known, as a resource for human geographers. The Net offers facilities across a wide and unstructured range. The number of specifically geographical areas is small although it is growing fast, and it would be too restrictive to concentrate on these few areas. Instead we will explore the range of resource-finding techniques. However, we will pay very little attention to some areas of Internet communication which have only marginal relevance to academic work or may not be made available in all universities. The possibility that the Internet might become the basis for a fundamental remodelling of the way we behave spatially cannot be neglected, but has to be flagged for the future rather than discussed in detail. Because the Internet is changing so fast, it can also be guaranteed that the description that follows will be out of date in detail by the time it reaches its first reader. However its basic principles and operational techniques are not likely to change in the next few years, and the topic is certainly too important to be ignored.

Students have a privileged place on the Internet as members of universities, and this text has been written on the assumption that its readers will have access to the Net through university links. Any student reader with an interest in establishing an independent home location can use one of the many technical guides available. The technical and commercial choices involved in becoming a free-standing Net user are not discussed here. As users of university systems students are typically users rather than active contributors to the Net, and this chapter reflects that emphasis. Once again, the guidebooks can redress this balance if necessary.

At this stage we have to review the basic workings of the system, before we can go on to look at the specifically geographical aspects of the Internet.

Development of the Internet

As a communications system, the Internet is based on the principle of networking. To participate in a network using current technology, a machine has to be physically linked to a communications line and must be host to software that will allow it to make use of the link. Networks began to emerge in the 1960s. Co-ordination and information-sharing became important issues as computers emerged from the specialised research laboratory and became routine tools in academic and commercial life, and this made networking attractive. One of the major tasks of networking is to subdivide an organisation's tasks between a number of specialised machines, a process that became more and more important historically as the evolution of computers shifted emphasis from large mainframes to workstations and personal computers in the

1970s and 1980s. In distributed computing of this kind machines are typically organised into short-range local area networks (LANs) within individual buildings or campuses.[1] The distributed machines might be used for specialised purposes, and the network will be administered by a central 'server' machine co-ordinating their behaviour, acting as host for software used by the other machines, and perhaps carrying out processing jobs beyond the power of the distributed computers. The significance of this from our point of view is that universities (among other organisations) typically have internal networks in place that can also be used as the first stage in linking individual machines to the outside world via the Internet.

Networking before the Internet was not simply a matter of local communication. Networks were also used to share information directly between individual machines or LANs that had previously been isolated or 'remote'[2] from each other, using dedicated machines known as 'routers'. Traditionally most localised networks were independent systems with relatively little need for external communication, and in fact the access allowed to outsiders would often be limited for security reasons. Open access was not usually encouraged, and until recently this kind of communication could be a difficult process needing knowledge of communications protocols and passwords that would give access to remote networks.

Conventionally the first step towards the Internet itself is seen as the creation of Arpanet as long ago as 1968. Arpanet was developed with the involvement of the United States Department of Defense, which had the aim of co-ordinating military and governmental information systems. To be at all valuable it would have to operate on a large scale, connecting computers at military, governmental, and scientific sites all over the continental United States and beyond. After a few years of development the designers made an important policy choice. They abandoned the concept of a single universal network in favour of technology that would provide a link between many networks of different types. This is the underlying logic of the Internet.

Throughout the 1980s the networking environment improved. Networks supporting communications in specialised academic and technical fields appeared in large numbers, and commercial information services were developed to serve particular markets using telephone-line connections. A separate development was the evolution of talk facilities and imaginary environments or MUDs (originally Multi-User Dungeons). The modern Internet detached itself from its military wing in the early 1980s and came to link and host communication systems of all these types as networking facilities improved. Quite suddenly in the early 1990s the Internet found a place in the public consciousness and developed explosively as individuals and small organisations realised that they too might have roles in this communication system. Like all new concepts, the Internet has been publicised in ways that are open to misinterpretation. For example the term 'Information Superhighway' has been widely used since it was popularised (and allegedly invented) by the American Vice-President Al Gore in the early 1990s, but it is a remarkably misleading image. Rather than feeding the individual user into the traffic on a giant multi-lane motorway, the Internet is much more like a traditional postal service, providing an international link between local distribution and collection systems with their own customs and traditions. Instead of driving along some frightening California freeway, the user posts and receives packages without having to be concerned about the route they take as they travel. The only key differences from a postal system are that the packages take very different forms, and the system can operate at a speed no postman could match!

Technically the Internet is based on a set of communications protocols (working rules) collectively known as TCP-IP (Transmission Control Protocol/Internet Protocol). TCP follows the well-established networking principle of breaking information down into identified 'packets'. TCP packet transmission allows a message to be broken down and interleaved among other traffic, rather than waiting for a block of space large enough for it to be transferred in its entirety. IP's main role is to select the best route available at any moment, and the different packets that make up a message might even be sent to their ultimate destination by different routes. When the packets reach their destination TCP takes over and reassembles them. Users impatiently watching a percentage counter at the bottom of the screen recording the gradual build-up of a file should console themselves with the thought that older less flexible protocols were much less efficient.

At this stage it is necessary to introduce a few more technical terms that will be used later. Although computers and physical networks underlie the Net, it is essentially software, and all programs on the Net can be classified as servers or clients. Servers are programs that provide or give access to resources of one kind or other, and clients are the programs that interrogate servers on your behalf. The clients are the programs that users deal with directly, and learning to use the Internet consists largely of finding out which clients give access to which resources, and how they can be used most effectively.

Computers on the Internet are usually known as hosts,[3] on the principle that they play host to resources that others might find useful. As a student

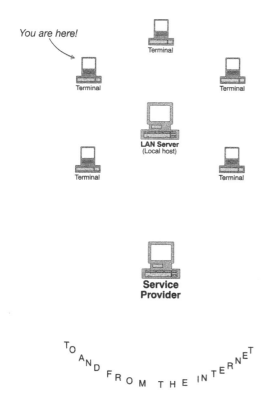

Figure 11:1 *The user's relationship to the Internet*

you are unlikely to work directly with a host of this type. It is more likely that you will use a networked computer that for Internet purposes is behaving as a terminal belonging to a host. Figure 11:1 shows how this arrangement might work and how it is linked to the Net. Whatever else it might be capable of doing, its Net role is simply to provide a link to a local host which is the machine that maintains the Internet link on behalf of its terminals. In technical terms universities typically offer undergraduate students network connections rather than full Internet access.

Every user on the Net has an address. The author's own address is *J.Lindsay@unl.ac.uk*. The first part is an alias for a User Identity or user ID. The part after the @ is the so-called domain name for the University of North London. Domains are aliases for the numerical codes used by servers navigating the Internet and they are organised hierarchically, rising higher in level towards the right. In the case of *unl.ac.uk* it is not difficult to work out that the University of North London is one member of the academic domain within the UK. Domain names usually follow general conventions. Education establishments are EDU in the United States (but AC in Britain). COM or CO identify commercial organisations, NET is used for networking sites, and there are various other suffixes like ORG, used for organisations which do not fit comfortably into the major categories. The domain type normally gives some prior evidence about the type of user you are about to access, although you should be prepared for almost any type of lunacy on some American EDU sites.

We have seen that the technology of the Net provides a communication system embracing a wide range of resources and types of information. If we take into account its rapid and unplanned growth over a short period, it is hardly surprising to find that there are several quite distinct areas of development within it, each with its own range of information, screen presentation, and group of users. Later in this chapter we will look selectively at some of the more important of these.

Getting started

Anyone with a computer can use the Net. There are no membership fees or annual charges. Indeed no organisation has overall control or ownership. Although there has been intense rivalry among major computer companies over the growing market for Net-related software and services, they are currently competing mainly in the way that they offer services within the agreed protocols. It is quite likely that at some stage in the future the winners of the contest might start to impose charges for services currently offered free. Of course even now there are no free lunches, and the individual user has to pay for access in one way or other. The costs usually arise from service provision and physical linkage. The normal way of linking a machine to the global network is via a so-called service provider, and these come in all sizes, from major corporations down to individuals running local services in remote areas. Service providers do all the work of maintaining contact with the Net and can also be used as hosts for an individual's home page or mail facilities, allowing the individual to be linked to the Internet on a 24-hour basis even if the home machine is switched off. Although a minority of service providers are run as free community facilities, the majority charge for the use of their services. Most provide little more than a network gateway, but some large providers had already developed their own subscription-based information services before the Internet blossomed, and are now coming to terms with the need to participate in the Net itself without losing their own user individuality. Compuserve is an excellent example.

The second charge that the individual user has to face is the telephone bill. To get access to the Internet an un-networked computer has to be connected via a telephone line using a modem. Although the telephone charge is based on the link to the service provider rather than the ultimate destination, extended use of a modem can still be expensive.

Students have advantages as Net novices, not least because the technical problems have already been sorted out by their host university and there are no direct charges. Most universities offer students individual access to the Internet via networked machines, and many also offer formal Internet classes or learning packages. The university itself will have negotiated its own contracts with service providers and will pay for these, and the student user does not have to worry about the cost of modems or large telephone bills. Universities have always been active members of the networking community and this expertise has made them natural centres of Internet activity. Academics still form a large group within the Internet community.

Being a student user is not an unmixed blessing. Not all universities offer the full range of facilities, and quality of access will vary. North American universities give a good deal of freedom to students on the Internet, and individual home pages run by graduate students have a world-wide reputation in some fields. The rather anarchic spirit of the Net and the speed with which it has been adopted on this side of the Atlantic have left the co-ordinating committees behind, and at the time of writing it was impossible to get more than an impressionistic idea of the range of services provided in European universities. The vast majority of universities and colleges in Britain are members of the Joint Academic Network (JANET), the long-established UK university network. Janet started operating in 1979 and developed an IP service (Janet IP Service or JIPS) in 1991.[4] UKERNA which currently operates Janet owns the physical linkages for the network, and its members are obliged to work within a Janet Acceptable Use code designed to limit dishonest, unethical, and frivolous use. UKERNA in fact recommends institutions to define their own Acceptable Use statements within its terms, and there is no general agreement so far among British universities about student access to the Internet.[5] The luckiest users might have 24-hour access and encouragement to develop their own Web home pages and other facilities. Others might find their contact time severely rationed and their access to facilities limited by cautious or restrictive local policies.

In general and rather bland terms it is fair to say that most universities will actively support student use that is academically relevant, particularly if it contributes to learning on particular courses. However it has to be recognised that huge amounts of Net energy are committed to topics that do not meet this criterion. Taking the most obvious case, the word 'sex' is said to be the most common keyword used in Net searches, and in his light-hearted survey of newsgroups Harley Hahn found that nine of the ten most popular among all Net users were explicitly sexual (Hahn 1996: 266–269). Students share the general enthusiasm for this topic, and while universities may accept this as inevitable they are likely to be concerned about use of the Net that might associate them directly or indirectly with abuses of sexual freedom, other problem areas such as financial dishonesty and political extremism, or indeed criticism of their own courses and administration. Since it is very difficult to monitor Net behaviour consistently, some universities have decided that the easiest way of maintaining control is to discourage or disable access to particular areas of Net activity such as Internet Relay Chat (IRC).

To use many of the facilities on the Net it is necessary to have a user ID. Although you may be able to send electronic mail using the generic address of a university network, you will need your own user ID to receive any messages in return. Any kind of interactive system like Usenet, which lets you 'subscribe' to different newsgroups, needs space to log and store your subscription information. This will not automatically be available on a student-access network. The host university may well be happy to give students mail and Usenet privileges but it will probably still be necessary to apply to the local administrator for a user ID. The level of access and the way it is administered will vary from place to place. You should ask your local computer service what is available.

The Web

The World Wide Web, usually now described simply as the Web, is one of the most recent developments within the Internet. It has grown so fast within its few years of existence that it has tended to eclipse the older service systems, and some users now regard the terms Internet and Web as synonymous. This is certainly not true, but as the Web has evolved it has become increasingly universal, providing a base from which the other services can be used. For that reason we will look at the Web first, and then go on to examine the less visible but still important areas of Gophers, mail, and file transfer.

The earlier forms of Internet communication were based on text screens, but the distinguishing feature of the Web is hypertext. The basis of hypertext is that it contains embedded links which allow the user to obtain other information or move to a different resource. Typically hypertext will be presented on the screen in a GUI environment like Microsoft Windows or Windows 95, with highlighted links that can be opened with a mouse click. One of the most exciting things about hypertext is that the links can open up graphics images, video clips, and sound samples as well as conventional text. Figure 11:2 shows a typical Web page. Links are shown by underlining and are also highlighted in colour in the original.

The first form of the Web developed from research carried out at the CERN Particle Physics Laboratory in Geneva during the 1980s. The World Wide Web project itself started in 1989 and a pioneer form came into use there in mid-1991, although it was initially intended only as an in-house application. However, it attracted external interest and was made available for public use on the Internet in early 1992. Because it was not available in an easily accessible form it tended to be confined to special

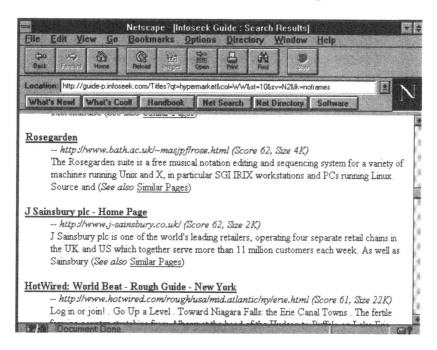

Figure 11:2 *Screen dump of a typical Web page*

interest groups, and the real breakthrough was the launch of Web-handling software capable of running in the widely accessible Microsoft Windows and Macintosh GUI environments late in 1993 (Hahn 1996: 165–171). It is a measure of the attractiveness of the Web that it has made so much progress in a space of less than three years at the time of writing.

Like other Net software the Web is founded on the server–client relationship, but the nature of the Web makes the relationship looser than in other forms. A single Web page might carry links to graphics or sound files, locally or remotely stored documents, mail addresses, or other Web sites. Each one of these must have its own unique address if the hypertext link is to be completed, and these addresses are known as Uniform Resource Locators or URLs. A typical URL (this one locates the home page of the University of North London) is *http://www.unl.ac.uk/unl/*. The '*http*' element shows that the URL contains information in the Web's Hypertext Transfer Protocol (HTTP) format, and the '*www*' element tells us that the address can be found by following a Web pathway. If the location belonged to another convention such as gopher or a commercial service like Compuserve, the first element of the URL would show this.

Web client programs are normally called 'browsers', and browsers allow the user to navigate through the Web. Broadly speaking a browser lets you carry out three important tasks. The first and most important one is to explore the resources available. Because there is no overall control over the Internet there is no central catalogue, and files are all the time being added or deleted without any centralised record being kept. The browser has to offer the user ways of searching this constantly changing pattern. If you have been given a particular URL address by a fellow-student, the browser will let you open this location directly by typing or pasting in the address. If you do not know where to look you can use one of many 'search engines' designed to search the contents of the Web at high speed for selected text elements. Search engines are capable of searching millions of documents for every occurrence of a specified word, phrase, or set of keywords. The output will display the first few lines of each matching document. Unfortunately the output selection is likely to contain large numbers of unwanted cases where the engine has found your text used in an irrelevant context. You will also find that even slight changes to the search text can produce quite different results. Only experience will let you get the best out of the process.

The second task is to keep track of the user's progress through a working session. Because it is uncharted, navigating the Web tends to be an erratic procedure. Even the most disciplined navigator might have to cope with numerous false trails during a search. Less dedicated searchers will find it very easy to be sidetracked into interesting but completely irrelevant areas. While a search is being carried out the browser will therefore keep track of the course it takes, and allow the searcher to go back to an earlier stage. This tracking information or 'history list' is lost at the end of a working session. The third important task is therefore to be able to select and mark useful URLs for later use. Browsers allow this to be done by bookmarking, which lets the user store the names of chosen URLs as links for later use. Bookmarking might not be enabled on networked common-user machines, and in such a case the student user will have to record accurately any URLs that might be useful later. This is common sense in any case.

Every hypertext URL defines the location of what is described in computer parlance as a 'file' of information, but life is made a little more confusing by the use of the word 'page'. A page is simply a hypertext file. Depending on its purpose and the owner's preferences, it might contain a few lines or enough information to fill the screen many times over. The term 'home page' refers to the main page for a particular site, whether it is a university, an airline, or an individual in Belgium with an interest in Mexican cookery. Many home pages carry links to subsidiary pages which of course will have their own URLs, and the term 'home page' is sometimes used in a looser sense to cover all of these, particularly for individually owned sites. In a slightly different sense, 'home page' refers to the page that is loaded first when you open your browser. It is likely to be the home page of your host university. If you want to build your own home page and can find a host to support it, you should use the Internet textbooks for information on techniques.

Depending on the nature of the page, browsers will offer different options for the user to capture their contents. Some links allow the formal downloading of files, usually in compressed forms that have to be unzipped for use. In other cases the file in its entirety might be saved to a local directory, or part of the contents can be highlighted, copied, and pasted into a suitable application. There is one important point to note. Hypertext pages on the Web are normally stored in Hypertext Markup Language (HTML) format. These are easily identified by having the htm or html suffix at the end of the URL. The browser reads the HTML file and interprets it to provide the graphic imagery designed by its originator. However if you find a file in HTML format full of interesting maps or other graphics and save it for local use, you will probably be disappointed. When the file is opened for reading in an editor or word-processor, you will find that the file consists of interleaved text and instructions. In place of every map or graphics item will be part of the URL[6] and instructions for processing. You might be able to retrieve items like this by expanding the URL to its full size, but this requires a little experience.

Because the Web is so diverse and rich, browsing sessions can be enjoyable but ultimately fruitless. In a few minutes of browsing, an unfocused user can get hotel booking information from New Guinea, download photographs of guinea pigs in Santa Claus outfits from their own home page in Hawaii or guinea pig recipes from Shanghai, map a route between stations on the Metro in Santiago in Chile or locate Chilean suppliers of PVC tubes, or even enjoy the collected speeches of the Governor of Alaska.[7] This kind of browsing can be entertaining, but it has to be said that among these examples only the Chilean underground case has any serious claim to be geographical.

However there are sources of genuine geographical value. Here is a selection that you might like to try, to get some idea of the potential of the Web. These should all be long-lived, but their URL addresses might change. If you can't find one of them, use a search engine to find its new location.

- *http://www.geog.le.ac.uk/cti/*   The CTI Centre for Geography, Geology, and Meteorology, the home page for the Geography area's British Computer Aided Learning project.

- *http://www.nene.ac.uk/aps/env/gene.html* GENE – the UK Geography 'Discipline Network' based at Nene College.
- *http://www.scit.wlv.ac.uk/ukinfo/uk.map.html* University of Wolverhampton UK Sensitive Map – Universities and Colleges. A comprehensive map-based index of British colleges, which lets you home in on virtually all UK academic sites.
- *http://www.esri.com/* The home page of ESRI, the specialist GIS house responsible for Arc/Info.
- *http://www.lib.utexas.edu: 80/Libs/PCL/Map_collection/Map_ collection.html* The Perry-Castañeda Library Map Collection at the University of Texas at Austin. This contains an interesting selection of downloadable maps. Be fair to everyone and do not try to download these at busy times.
- *http://www.bodley.ox.ac.uk/nnj/mapcase.htm* The Map Case – Bodleian Library Map Room. More downloadable maps, this time with a British historical focus.
- *http://www.ordsvy.govt.uk/* The Ordnance Survey – the primary British mapping agency.
- *http://www.geog.buffalo.edu/GIAL/netgeog.html* A page of geographical links from the Geographical Information and Analysis Laboratory of the University at Buffalo, NY.

At the time of writing the main geographical societies in Britain had not discovered the Internet.

You will probably want to build a list of sites suited to your own particular interests. Search engines are important for this, once you recognise their limitations. For the purposes of this text the author carried out a quick search for retail keywords. Using the word 'Superstore' seemed to be a reasonable start, but the attempt was a failure. The output was dominated by pages providing virtual services on the Web, which had adopted the word as part of their own title. A search for 'Hypermarket' was more successful, and the search soon turned up a major store chain with a home page including maps and lists of store locations. However as Figure 11:2 shows, it was keeping rather odd company linked only by having the word 'hypermarket' somewhere in their texts. It was also a single case, and some other large retailers have not yet become Web users. Text searching with an engine is not a 'smart' process, and it is not possible to make a single-pass search which will find all the sites which share a theme and eliminate the irrelevant. In fact searching the Web is far from being the easy process it might seem at first, and even quite small differences in the keywords used and their sequence can have surprising effects on the output of the search.

Exciting as the Web is, it has weaknesses. It has to be said that it contains a great deal of clutter, and the pressure of use created by its success also slows it down at key times. It is a very good idea to carry out any Web work as early as possible in the day, while most of the North American users are still asleep. It can be very slow and frustrating to navigate the Web during the evening in Europe, when users all over North America are awake and active on the Net. The absence of central control or agreement means that it is not consistent. Different organisations in the same field may provide information completely different in type and quality. It lacks large-scale systematic sources. The Web is not a place to go directly for bibliographies or reading lists, and large-scale data files of any value are very rare. Few organisations put information of this kind in the public domain, although the Web at least provides mail addresses which can be used to open contact with them. Some excellent resources on the Web are also provided by organisations or individuals maintaining pages which consist mainly of links within specialised fields.

Gophers and gopherspace

Gopherspace is the domain of the gopher. The principle of the gopher is another old one, used widely in local contexts under different names before the development of the Internet. You should think of the computer gopher as a fetcher and carrier, like its human counterpart in American slang – the 'gofer'. The Internet gopher system was developed in 1991 as a way of linking local information networks, and had a flourishing but short career before being eclipsed by the Web two years later (Hahn 1996: 317). The development site at the University of Minnesota – 'the mother of all gophers' – still has an important role in gopherspace. Unlike the hypertext of the Web, gopherspace is text space and is driven by menus. The menus might offer a choice of locations for connection or services at a particular point. The client program works with a gopher server to offer successive menus to the user, according to the choice made at any level. Because it is text-based the gopher system is usually faster than the Web. Because gophers are much more firmly tied to systematic knowledge than the Web and lacks its graphic attractions, gopherspace is not full of self-indulgent home pages and other 'noise'.

Although further large-scale gopher development has probably been inhibited by the arrival of the Web, gophers still offer an efficient way of working the Net in some contexts. Sometimes you will enter gopherspace painlessly by following a link in a Web. Gopher clients offer some of the same navigational services as Web browsers, including history lists and

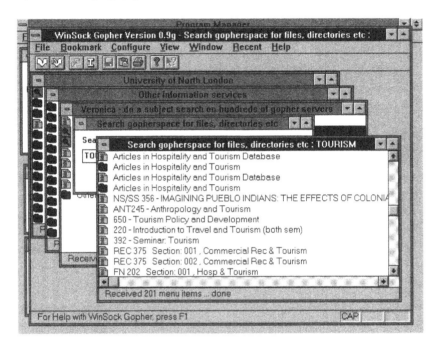

Figure 11:3 *A Veronica search in progress*

bookmarking. The contents of gopherspace are very varied, and navigating blind can be both frustrating and inefficient. If you know what you want but not where to find it the best course is to use the Veronica search engine. Veronica developed from a search engine (called Archie) designed for anonymous ftp (see below, p. 182) and is essentially a keyword search. At regular intervals Veronica trawls all gopher menu titles and updates its keyword files. Searching then becomes simply a matter of asking Veronica to find text items or 'strings' and using the menus that will be generated in response. Files can be downloaded quite easily. Figure 11:3 shows a Veronica search in progress, this time searching for information about tourism. Like Web search engines, Veronica can produce startlingly different results from searches based on slightly different wording. There are also differences between the results produced by working with different Veronica sites, and only experience will allow you to get the best out of the search procedure.

Mail

The idea of electronic mail (email) is old in computing terms. Long before the Net appeared on the scene, sophisticated mail systems already existed with facilities for document and image enclosures as well as simple text. The Net has taken existing forms and built mail carriage into the TCP/IP protocols mentioned earlier. In doing so it has made mail more accessible by providing more attractive mail environments. In theory at least, messages can be sent world-wide more or less instantaneously, although it can take a surprisingly long time for a message to be routed to its destination. The Web has made mail sending much easier, particularly where addresses are defined by hypertext links, and if you are lucky you might find that some helpful person has built the links for a series of related addresses into a Web page. As you browse the Web you will be aware of the range of static pages you view, but may not appreciate the size of the currents of mail which are passing through the Net all the time.

Email is not perfect. No matter how fast it travels, the message may sit at its destination for days before it is read, and longer still before it is answered. It might be possible to send messages from an open university network machine but to be a full mail user it is necessary to have a recognised address of your own to receive incoming messages. You may have to ask your system administrator for advice about this. The email system is not free of junk and abusive messages and if you are not careful in subscribing to mailing lists your local space might be flooded with unwanted files. You will also need the mail address of the recipient. One of the big difficulties about email is that there are no universal directories of addresses. Your host university will probably have its own on-line directory of internal email addresses, but to send mail to a remote site there has to be some means of finding the address. If you have not been able to obtain it from its owner or a link in a Web page, it might be possible to search for the address by asking the WHOIS service to look for an element you think will be part of it.[8] If none of these aids is available you will probably be unlucky. Guesswork does not go very far with mail addresses, and TCP/IP lacks the lateral thinking skills of your local sorting office.

Telnet, anonymous ftp, and other file transfer techniques

As the Net has evolved, so have methods of finding and capturing information. This text will pay much less attention to them than to the Web, for two reasons. The first is that they are still relatively difficult for novices. At one stage they were handled almost exclusively by command line interface, and the user had to master sets of unintuitive commands. The UNIX environment in which most of them were embedded was (and still largely is) famously unfriendly to the new user. As time has gone on service providers have tried to build more attractive environments for the older forms, but it has to be said that these can still be daunting for the uninitiated. The second reason for paying less attention to these techniques is that the development of the Web has made it much less important to know about them, either because it incorporates them but uses them more or less invisibly, or because it provides alternatives. The Web worker may unwittingly be using these older techniques quite regularly. The most important ones are described here, but readers who want to use them outside the Web context should read the standard guidebooks and get as much advice as possible from their computer service about the way they are set up locally.

Telnet is a UNIX-based application, and the oldest of the Internet communication modes. It allows you to log on to a remote computer. If you are working on one computer but have an account on a different computer telnet lets you log on with your own password as though you were on its own site. If you have no account or user ID on the remote host, most areas will be closed to you because you do not have the password needed to open the gate, but there are public resources accessible without a password. Finding these in the wide open spaces of the Internet is the first challenge. Unless you have been given the address of the files you want, you will have to use a search technique like Archie to locate and retrieve them. Because of its great age in Internet terms telnet is rather unfriendly. Although telnet can be used to handle graphics files, it is firmly text-based with its own special language and conventions. It is also important to be aware that it is essentially an enabling device. Once it has let you log on to a remote machine, the rest is up to you. Although client programs have been developed for Windows and other GUI environments they tend to ease only the first contact, and the intending user will still need to have some knowledge of telnet conventions. If in doubt, consult a text such as Hahn (Hahn 1996: 469–487).

Anonymous ftp is the rather puzzling title for one of the more useful elements of the Internet. Long before the Internet was invented, computer users were transferring files from one site to another using the File Transfer Protocol or ftp. As with telnet, security means that it is not usually possible to access a user area on a remote host without logging on with a user ID and valid password, and this obviously closes the door firmly in the face of casual browsers. The significance of anonymous ftp is that it permits general access to designated areas containing files that can be freely downloaded. Ftp sites are like Internet public libraries, the one difference being that anything you choose to borrow does not have to be returned. It is still necessary to log on, but this is done by using 'anonymous' when asked for a name. Your full mail address should be used when you are asked for a password (if you are lucky your client programs will help by providing a fast route which automatically finds and uses your mail address). The significance of the mail address should be clear enough. It provides ftp with the address it needs if files are to be transferred.

Once inside the remote host, your task is to navigate through the directories available to you searching for suitable files. This is quite easy work with the mouse if you are working through a GUI-based client program, but if you are working in a command-line environment you will have to know the techniques of navigation in the host area. When you have located the file you have to choose appropriate settings. If you know that the file consists of straightforward text you should ask for it to be transferred in ASCII format. If it consists of graphics, executable software, or is in compressed form you should ask for binary transfer. When that is done, the file can be transferred quite simply to your own machine and is ready for use. If you are using a good client program, it will probably have its own list of important ftp sites and the facility of letting you add your own choices to this list.

Ftp sites are typically big and institutional. Universities and research organisations dominate the scene. One of the most important and widest ranging in the UK is Imperial College London at *doc.ic.ac.uk*, which not only has large resources of its own but has 'mirroring' arrangements with other major sites designed to spread the load at busy times. Not all academic subject areas are equally represented in ftp sites, and the evolution of the Internet from the specialist community is reflected in the continued importance of Computer Science areas within them. However, there are also news and other services, and sites which have non-academic specialist interest.

All areas of the Net are busy, and anonymous ftp is no exception. Because every ftp login adds to local congestion within the remote host network, it can take a long time to get a connection. Ftp hosts implement their own policies about external access. Some will forbid all anonymous access at key times and others ration access by allowing only a defined number of visitors access at any time. The best policy is to concentrate your ftp searches at times when the Internet as a whole is quiet (i.e. when the majority of Americans are asleep), but more particularly outside the normal office hours of the site you are visiting.

Talk and interaction

Some users find that the main attraction of the Internet is the range of interpersonal links it offers, in the form of talk and chat services, special interest groups to which they can contribute, or shared environments in the MUD mode. There is not a major place for these here. By their nature these facilities have only a limited contribution to make to academic fields. Pressure on resources and concern about misuse mean that system administrators in university networks may not offer some of these facilities to student users in any case, and this author does not want to make their lives more difficult by encouraging students to lobby for the use of MUDs and chat services.

The one area which should not be neglected is Usenet. This is the system that houses Newsgroups, Internet areas in which individuals can post and respond to messages on specialised subjects. All newsgroup names start with a diagnostic element, so that SCI and SOC indicate respectively scientific and sociological groups. REC identifies recreational groups and there are regional identifiers like UK. One of the largest and most varied groups is ALT (Alternative). Usenet is the Net at its best and worst. There are serious newsgroups like uk.environment and alt.energy.renewable, but they are heavily outnumbered by special interest groups like alt.music.genesis, alt.alien.visitors, and even rec.collecting.stamps. Inevitably there are large numbers of groups on the wilder shores of sexuality and others which are no more than juvenile ranting.

Hahn (1996) identified and categorised about 5,000 newsgroups, but the number is constantly changing and the real total is probably a lot higher. However, his survey probably reflects the general balance reasonably well. Broadly speaking about 40 per cent of the newsgroups identified by Hahn were 'mainstream', and most Usenet sites will carry a large number of these. Unfortunately few of these are directly relevant to geographers. The other 60 per cent were 'alternative' (the ALT category itself contained over 50 per cent of the global total), and site administrators may choose to exclude some of these (Hahn 1996: 260–261). On balance, academic life is served by only a small proportion of newsgroups – if we exclude computing itself, fewer than 10 per cent of the total are mainstream groups in areas of academic interest – but they have a good chance of being available. Your university will have a link with a particular Usenet server, and its policy will decide what is available. A site administrator with a sense of moral responsibility may deny you access to groups like alt.sex.fetish.watersports, and if disk space is short groups like alt.elvis.sightings may be excluded as well, but the worthier sites are likely to survive.

Like other long-established Internet areas, newsgroups can be accessed by command line or GUI client programs. The normal procedure is to 'subscribe' to interesting groups. This opens up the possibility of browsing their contents and contributing if appropriate. At a later stage the group can be 'unsubscribed'.

Referencing the Internet

The Internet has created new opportunities for the academic community, but it has also created new problems.

One of the biggest problems is the verification of Internet information. Printed sources have well-established mechanisms of quality control. Publishers of academic books and journals use a system of readers and referees designed to prevent carelessness or dishonesty marring the books they publish. This does not guarantee that every text or article published in the field of human geography is factually accurate in every detail and that all readers will find its analysis acceptable, but it does impose a consistent discipline. Electronic publishing is not intrinsically careless, but the speed of growth in the Internet has outstripped the established process, and the results are not always satisfactory. There may be ambiguity over authorship, sources, and citations, and inaccuracy in proof-reading or data presentation is not uncommon.

Electronic sources are also unstable over time. Printed sources almost always carry a publication date. Even if a book goes through several revisions and new editions, it should be clear where a particular copy belongs in this publication history. Computer-generated files lack this permanence. The owner of a file available on the Internet has the power to modify or delete it at will. If you download a copy for use in preparation of an essay or dissertation, a reader following up the reference later might not find it at all or might find a significantly different document under the same name. The originator of the file may know its revision history, but this is very seldom made clear in the text of the file itself. Many Internet sites are short-lived, and a later reader might not be able to access the location of the file at all.

Corruption and deception are also dangers. Malicious counterfeiting of files is not unknown, and bogus files are certainly much easier to produce than forgeries of printed sources. Although inspection of the URL will normally let you assess whether or not a file comes from its stated source, it does not give an absolute guarantee. Late in 1996 the British Labour Party home page was sabotaged by hackers who altered the text in very visible ways and substituted cartoon images for photographs. High-profile action like this will be spotted instantly, and in this case the perpetrators for their own reasons made sure that it was well publicised, but small mischievous alterations are easy to carry out and might be very difficult to detect. It would take only seconds for a prankster in a shared office to change a few vital words in an open file on an unattended machine.

How do we cope with these problems? Suppose that you are visiting enthusiasts' home pages to get information about images of places in Hollywood movies, or storing a sequence of store locations maps produced by a retail group over time, or assembling information about a contentious issue like a by-pass scheme. All of these routine processes might involve one or more of the problems mentioned above. Eventually the academic community will evolve a set of rules to handle Internet sources, but in the meantime we have to improvise. Here are some suggestions based on local practice in the author's home Department.

- Take care! Always evaluate an electronic source carefully.
- Use electronic sources only where there is no reliable printed alternative. This is particularly important for major references.
- As far as possible use electronic sources only where they come directly from organisations with public or academic reputations to protect.
- Make a careful note of authorship and version history. If possible download or copy the document and keep the file secure. Even if there is no visible version history, a properly downloaded file will normally be invisibly 'stamped' with a date and time of creation, and this can be inspected by command line directory listing or use of a file manager. If you cut and pasted to produce your version of the document it will be stamped with the date when you did this.
- Cite electronic sources as you would cite published material. Include the name of the source organisation or individual, and the URL or document location.

Summary

Many geographers and other academics mistrust the Internet. The strongest criticisms usually come from those who have never used it but are aware of its reputation as the home of computer nerds and child pornography. Others have experimented with it and been alarmed by its anarchy, its untidiness, and lack of academic formality. These apprehensions have some grounds, but the potential of the Internet is far too great to be ignored. It is almost inevitable that during the next few years the Net will leave its pioneer phase. How it stabilises then will depend very much on the wishes of its users. If geographers and other academics are not actively involved, they will lose the chance of moulding its future form.

Discussion questions and tasks

1 Can the Internet really create a spaceless geography? Think about the implications of this question.

2 The Net has been hailed as a radical means of empowering people through education and reducing the world's inequalities. It is also big business. Key current players include Microsoft (the founder and Chairman of which is Bill Gates, the world's richest man) and Netscape (a Web browser house which has become a multi-billion-dollar corporation in less than two years of existence). Can global equalisation and global commerce coexist?

3 Carry out a Gopher search for bibliographies in a field that interests you, and retrieve your findings.

4 Carry out a Web search for sites providing information about one of the following topics: tourism in Portugal; BSE; urban graffiti; world tanker fleets; the return of tuberculosis. Keep a record of your search methods and paths.

Further reading

See also

Data sources, Chapter 3
Computer techniques, Chapter 8
Computer mapping, Chapter 9

General further reading

Hahn, H. (1996) *The Internet: Complete Reference*, Berkeley: McGraw-Hill, 2nd edition.

The title describes it well. This text is one of the best general-purpose introductions to the Net.

Krol, E. (1994) *The Whole Internet User's Guide and Catalog*, Sebastopol, CA: O'Reilly, 2nd edition.

Another of the more interesting general texts, although this second edition is inevitably less up to date than Hahn's.

Pope, I. (1995) *Internet UK*, Hemel Hempstead: Prentice Hall International.

Quite a useful local text although much of it consists of lists of sites, and a lot of these have become outdated even in the short time since publication.

Schofield, S. (1995) *The UK Internet Book*, Wokingham: Addison Wesley.

An entertainingly written and fairly comprehensive guide to the Internet from a UK perspective.

You can improve your Web expertise and explore the range of texts by using Net-based bibliographies. The best is probably Kevin Savetz's enormous 'Unofficial Internet Booklist'. The latest version (June 1996) is at *http://red-wood.northcoast.com/savetz/booklist/*. Avoid older versions at different locations.

Notes

1   Networks can be members of networks. A university (for example) might operate a Wide Area Network or WAN linking LANs in different Faculties, Schools, or Departments.

2   The term 'remote' has nothing to do with physical distance. Two machines sitting on the same desk but belonging to different networks are remote from each other in this sense, and no less so than if one was on a desk in Australia. The great strength of the Net is its potential to make the machine in Australia no more remote than the one on the same desk.

3   An older but still current use of the term 'host' is to describe a timesharing computer supporting a set of terminals. The terms can overlap. It is quite possible that as a student user you will work on a terminal which is linked to a machine which is both a timesharing and Internet host.

4   A Web history of Janet can be found at *http://www.ja.net/jnt-history.html*

5   The Janet Acceptable Use Policy had reached Version 4.0 by April 1995. Copies can be obtained from UKERNA on paper or on the Web from *http://www.ja.net*. UKERNA officers occasionally use the Net to complain about resource wasting and other frivolous use. If you have any doubts about the terms on which you are using the Internet from a university site, you should consult your local Computer Service.

6   Usually the later part, needed to find the image from the host URL's location.

7   Readers who do not believe this can practise their URL location skills on them. The New Guinea site is *http://www.hotelstravel.com/papua.html*. The pet guinea pigs are at *http://www.hawaii.edu/randomstuff/gpigs/* and the edible ones at *http://www.sh.com/dish/dish079.htm*. The Santiago Metro is *http://metro.jussieu.fr: 10001/bin/select/english/chile/santiago-de-chile* and the Chilean Yellow Pages at *http://www.chilnet.cl/rubros/TUBOS10.HTM*. Finally the Alaskan rhetoric is at *http://www.gov.state.ak.us/local/gov/speech/ table.html*. The Web changes fast, and some of these may have vanished by the time you read this text.

8   WHOIS can be accessed through a number of servers. WHOIS searches tend to pick out active Internet users and may not identify addresses for people who seldom venture out of their own environment. For help in this field see Schofield (1995: 62–63) or Pope (1995: 65–69).

Glossary

Address In computing, the unique identifier which gives the user access to **email** and **Internet** services. It includes identities for the user and the **host** machine.

Alphanumeric data **Data** recorded in textual form. They cannot be used directly for numerical analysis.

Alternative hypothesis In hypothesis testing, the **hypothesis** which can be accepted if the **null hypothesis** is rejected.

Anonymised data **Data** processed to ensure that the user cannot identify the individuals or organisations to which they relate. Typically necessary where very few cases fall into some categories.

Anonymous ftp An elderly but still useful mechanism for accessing **remote** sites on the **Internet** and transferring files. FTP stands for File Transfer Protocol.

Attribute In spatial computing, a **data** item belonging to an **entity**.

Autocorrelation An arithmetic relationship between groups of **data** items which might be selected as part of a **sample**. Thus the percentage of males in a **population** directly affects the percentage of females. Autocorrelated data cannot be regarded as statistically **independent**. See **spatial autocorrelation**.

Before-and-after survey A survey which revisits the same **respondents** before and after a particular event.

Bitmap An image displayed by mapping values to the individual **pixels** on a screen from a stored file.

Browser **Client** software designed to let the user navigate and retrieve information from the **World Wide Web**.

CAD Computer Aided Design. A generic title for graphics software used in engineering design, architecture, and mapping.

Cartesian grid A Cartesian grid is square, and any location has unique **co-ordinates** defined by recording its position horizontally or west-to-east (X axis) and vertically or south-to-north (Y axis).

Cartography The science of map production and analysis.

Case In statistics, an item about which information has been obtained. The term **individual** is sometimes used with the same meaning.

CD-ROM Compact Disk/Read-Only Memory. A high-volume, robust, portable computer **storage** disk, increasingly used in **hypertext** applications.

Central tendency Measures of central tendency are **descriptive statistics** (like the mean) that define a form of centre around which the set of data can be supposed to be distributed.

Chart A map designed for air or sea navigation.

Chi-square test A **non-parametric test of difference** for two or more **independent samples**.

Client **Internet** software that lets the user interrogate a **server**. It may be specialised or offer a general search facility.

Clip-art A prepared image designed to be inserted into a computer-generated document. Usually obtained from file libraries.

Closed question In **interviews** or **questionnaires**, a question for which the answer must fall within a defined range.

Cluster area sampling A method in which random selection is used to identify units of a hierarchical **sampling frame** from which **respondents** will be drawn.

Co-ordinates Numerical values recording positions in space. See **Grid**.

Confidence level The level of **probability** at which it can be said that a **parameter** falls within a given numerical range.

Confidence limits The values which bound the range defined by the **confidence level**.

Coverage A term used in vector-based **Geographical Information Systems** to describe a stored set of **entities** of a particular type.

Cross-section survey A survey which treats all its **data** as being obtained at the same time. The term snapshot is sometimes used.

Data Items of information appropriate for a particular **research** task.

Data generation The process by which **data** are obtained. Several other terms are used. See the text for a discussion of these.

Database A systematically organised store of **data**.

DBMS Database Management System. Software designed for storage and manipulation of **databases**.

Deduction The method of inference which sets up a series of premises and conclusions as a logically consistent way of providing general explanations. These can then be tested by application to real world cases.

Degrees of freedom A number which allows **sample** size to be taken into account in assessing the **significance** of a **test statistic**. Degrees of freedom vary from test to test.

Depth interview An informal **interview** giving the interviewer the ability to vary the depth of questioning on different topics as appropriate.

Descriptive statistics Those which in themselves do no more than summarise the characteristics of **data** sets. See **inferential statistics**.

Desktop publishing Software designed to let the user create, edit, and present text-based documents to a higher standard than **word-processors** allow.

Digital map A map created and stored using one or other form of computer technique.

Digitiser Or digitising tablet. A device to let the user transfer patterns from maps to digital **storage** by identifying and storing key sets of **co-ordinates**.

Digitising The conversion of map patterns into digital information that computers can store, read, and display. Usually associated with **vector data**.

Diskette The proper name for the small-format disk or 'floppy' form of data storage.

Dispersion Measures of dispersion are descriptive statistics describing the ways in which values are spread through the **data** set.

Distribution See **probability distribution**.

Earth spherical co-ordinate system The grid of latitude and longitude lines that define positions on the earth's surface. Lines of latitude run parallel to the equator. Lines of longitude (meridians) run from pole to pole, and their separation varies with the curvature of the earth's surface.

Email Mail communication using computer **networks**.

Entity In **vector**-based graphics, an **object** composed of a one or more lines, points, or areas. Typically entities will form the spatial skeleton of a map, and will have their own **attributes**.

Error In **hypothesis** testing, rejection of the **null hypothesis** when it should be accepted is a Type I error, and acceptance when it should be rejected is a Type II error.

Ethnography The study of people's behaviour in social contexts, with a particular emphasis on interaction in everyday situations.

Extensive research Research based on taxonomic groups, classifying individuals on the basis of shared attributes rather than examining the way in which they interact. See **positivism, quantitative**.

Factual question In **interviews** or **questionnaires**, a question calling for a factual response which could in theory be checked by independent observation.

Falsificationism The doctrine that since no **hypothesis** can ever be fully verified, science progresses by accepting hypotheses provisionally until they can be proved false and therefore rejected.

Filter question In **interviews** or **questionnaires**, a question which directs further questioning into a particular path according to the response.

Focus group A group of people assembled for a moderated discussion concentrating on a particular topic.

Geographical Information System Software which allows the storage, display, and analysis of spatial patterns and sets of **data** associated with them.

Gopher One of a family of **client** programs for searching and retrieval on the **Internet** based on use of a hierarchy of text-based menus.

Graphicacy Competence with graphic techniques.

Grid A mechanism for recording **data** positions. See **Earth spherical co-ordinate system** and **Cartesian grid**.

GUI Graphic User Interface. A computer interface designed to be user-friendly, with an emphasis on images and links.

Host A term with several meanings in computing. Two are important here. In a multi-user networked system the host is the machine which supports a group of terminals. A computer with a direct **Internet** connection is also called a host. Many university computers are hosts in both senses.

Hypertext Text containing embedded links which give access to information stored elsewhere. A key element in the **World Wide Web**.

Hypothesis A testable proposition in a research context.

Hypothetico-deductive method The procedure by which research progresses through setting up successive **hypotheses** which are then tested against observations to see whether or not they can be disproved. See **falsificationism**.

Idiographic Concerned with exploring individual connections. See **nomothetic**.

Independence A requirement of **data** used in statistical testing. The most important criterion is that they should be free of **autocorrelation**.

Individual See **case**.

Induction The method of inference in which observation of individual **cases** is used in the formulation of **hypotheses** that will solve general questions.

Inferential statistics Tests which allow assessment of **probabilities** and can thus be used in **hypothesis** testing.

Intensive research Research based on the relationship between individual cases rather than the study of aggregate classes. See **ethnography, interpretivism, qualitative**.

Internet The informal global **network** supported by **TCP/IP**. The ultimate **WAN**.

Interpretivism A methodology which treats human behaviour and its manifestations as **texts** to be analysed.

Intersubjectivity The relationship between the subjective behaviour of researcher and **participant**.

Interview A dialogue between researcher and a **participant** in a study. It may range from formal and structured to free-ranging and open, according to research context.

Janet Joint Academic Network. The UK's academic communications network, and owner of the physical linkages which support a substantial part of the **Internet** traffic in the UK.

Kurtosis The degree of concentration of values in part of a frequency distribution. If values are highly concentrated locally the distribution will have a marked peak (leptokurtosis). Weak concentration produces a flattened curve (platykurtosis).

LAN Local Area Network. Typically a localised **network** set up to meet an organisation's internal needs, with dedicated cable linking the members.

Longitudinal survey A survey in which individual respondents are revisited at intervals over a long time-span.

Mann-Whitney U test A **non-parametric test of difference** for two **independent samples**.

Measure of association A **test** designed to assess the strength of the relationship between two **data** sets.

Memory In computing, an area of **storage** dedicated to short-term holding of information for immediate use.

Methodology A general approach to the study of **research problems**.

Modem Communication device allowing independent computers to be linked to the **Internet** via telephone lines.

Net See **Internet**.

Network Two or more computers linked so that communication is possible.

Newsgroup A discussion and information exchange area on the **Internet** dedicated to a particular area of interest.

Nominal data **Data** consisting of items which are categorised only as types within a series.

Nomothetic Concerned with establishing generalisations and laws. See **idiographic**.

Non-factual question In **interviews** or **questionnaires**, a question which explores the respondent's attitudes, opinions, or beliefs.

Non-parametric statistics **Tests** which do not assume a particular distribution in the populations they examine. Equivalent to **distribution-free**.

Normal curve The bell-like curve that represents the continuously distributed values of the **normal distribution**.

Normal distribution Properly, the normal probability distribution. One of the most important **probability distributions** in statistics, represented by the **normal curve**.

Null hypothesis In hypothesis testing, the **hypothesis** (symbolised as H_0) which proposes that the relationship evident from a **sample** set of **data** does not truly represent a relationship in the whole **population**. Rejection of the null hypothesis will allow the acceptance of the **alternative hypothesis**.

Numeric data Data recorded in numerical form.

Object A term with several meanings in computing. In the graphics context it is sometimes used as an alternative to **entity**.

Objectivity The notional ability of the researcher to pursue pure **research** uncontaminated by personal value judgements.

Open question In **interviews** or **questionnaires**, a question which allows the respondent to answer freely.

Ordinal data **Data** which can be ranked or assigned to a place in a series.

Paradigm A model which provides a general framework within which a **research** community operates.

Parametric statistics **Tests** which are based on assumptions about distributions in the **populations** they examine, particularly that the populations approximate to the **normal distribution**. They are generally more **powerful** than non-parametric tests.

Participant observation A method in which the essential feature is that the researcher studies a community while living within it.

Participant A person providing **data** in a research exercise.

Pearson's product–moment correlation A **parametric measure of association** between two **samples**.

Pilot A preliminary study designed to test the practical soundness and validity of a research task.

Pixel the smallest individual unit of a screen display. The term is derived from the term 'picture element'. There will typically be hundreds of thousands of pixels in a display.

Plan A detailed large-scale map. The Ordnance Survey defines its 1:2,500 and 1:1,250 maps as plans.

Platform In computing, the type of machine available as a host for software. Convergence in design means that choice of platform is much less important to users than even a few years ago.

Poisson distribution A **probability distribution** appropriate in the analysis of series of discrete events.

Population In statistics, the theoretical total number of instances from which measures of a particular value could be derived. Statistical tests are usually applied to **samples** from populations.

Positionality An emphasis on social context or position within a group.

Positivism The **methodology** which seeks to find laws or regularities using **quantitative** methods.

Power The ability of a statistical **test** to make a correct distinction between a true and false hypothesis. Some tests are intrinsically more powerful than others, and most become more powerful when applied to larger populations.

Probability distribution The distribution of all possible outcomes of a situation and their **probabilities**.

Probability The likelihood that something will occur. A great deal of emphasis is put in statistics on ways of measuring the probability that an event will take place within a given number of occurrences, and assessing its **significance**.

Project design The process by which a research task developed using a **research design** is implemented on a day-to-day basis.

Projection A means of systematically displaying the curved surface of the earth on a flat surface.

Qualitative concerned with the meaning and value intrinsic in individual cases rather than **quantitative** generalisation from individuals to large **populations**.

Quantitative Based on the collection and analysis of numerical **data**.

Questionnaire A set of questions prepared in advance for a participant to answer. Questionnaires may be administered postally, by telephone, face to face, or by the participants themselves.

Random sampling A method in which a group of **respondents** is drawn at random from the **sampling frame**.

Raster scan A screen display in which the image is built up by assigning colour and tonal values to the individual **pixels** from which it is built.

Ratio data **Data** based on a scale which is defined in terms of equal units, and having an absolute or natural zero point.

Reliability The consistency with which **data** can be recorded. Reliability does not guarantee accuracy or appropriateness.

Remote In computing, describes a machine that is not part of one's local **network**. Physical distance is not a criterion.

Replication In laboratory science, the verification of research results by independently repeating the experiment.

Representativeness A primary aim of the **sampling** process is that the sample should be representative of the **population** from which it is drawn.

Research design The strategic process by which a **research** task is conceived, put into operation, and completed.

Research problem The question which a **research** task sets out to answer. For obvious reasons the term 'research question' is also sometimes used.

Research Work carried out with the aim of answering questions in any of the sciences.

Respondent A person or organisation taking part in an **interview** or **questionnaire** study.

Response rate The proportion of targeted **respondents** who co-operate with the researcher.

Rigour Strictness or discipline in the handling of research. See **objectivity**.

Sample A group of **cases** selected as a surrogate for the whole **population**. The sample should be **representative**.

Sampling error The calculated range within which the results obtained from a **sample** are likely to differ from those we would have got by using the whole **population**.

Sampling fraction The proportion of the **population** (if it can be exactly determined) represented by the **sample**.

Sampling frame A list or structure used to selected **cases** from a **population** to form the **sample**.

Scale The relationship between dimensions on a map and the real-life dimensions of the objects the map represents.

Scanner A device to convert images into digital form by breaking them down into small component elements which can be mapped onto the **pixels** which create a **raster scan** image.

Semiotics The study of the language of signs.

Server (i) A computer which co-ordinates and provides resources for a **network**. (ii) On the **Internet**, the hardware/software set-up at a particular location that provides resources for **client** users.

Service provider An agency (normally commercial) providing access to the **Internet**.

Significance Statistical significance is the level of **confidence** with which we can accept that an observed relationship or difference between **samples** can be taken as meaningful rather than the result of chance.

Skewness The degree of deviation from symmetry about the mean in a frequency distribution.

Spatial autocorrelation The situation in which values taken from locations close together are more strongly related than those separated by greater distances. See **autocorrelation**.

Spearman's Rank correlation coefficient A **non-parametric measure of association** between two **samples**.

Spreadsheet Software for numerical handling, in which data are stored in individual cells which can be linked to allow automatic updating and the use of formulae.

Standard deviation An important measure of **dispersion**, the square root of the **variance**. In a normally distributed population known proportions of occurrences can be expected to occur within different multiples of the standard deviation on either side of the mean.

Storage In computing, means of long-term holding of **data**. Most storage still uses magnetic media such as tapes and disks.

Stratified sampling A method in which the **sampling frame** is subdivided spatially or otherwise into sub-units which are then sampled.

Structuralism An approach emphasising a rule-governed model of social behaviour and interpreting individual behaviour in terms of social structures.

Subjectivity The degree to which the behaviour of the researcher (or anyone else) is affected by personal values and attitudes.

Symbolisation The use of symbols to convey meaning on a map.

Systematic sampling A method in which **respondents** are selected from the **sampling frame** on a structured basis.

T test A **parametric test** of the difference between two **samples**.

TCP-IP Transmission Control Protocol/Internet Protocol. The family of network communication protocols (standards) on which **Internet** communication is based.

Technique A working method used in the study of a **research question** within a particular methodology.

Telnet A long-established mechanism for accessing user areas on **remote** computer sites.

Test statistic The value derived from the application of a statistical **test**.

Test Or statistical test. A recognised means of establishing a statistical relationship and assessing its **significance**.

Text In the context of **interpretivism**, the term can be applied widely, including behavioural patterns, graphic and filmic data, and landscapes as well as written documents.

Thematic map A map designed to show patterns relating to a specific topic, often abstract.

Theoretical sampling A term sometimes used to describe the method of selecting for **qualitative interviewing** people who have distinct and important perspectives on the theme of the research question.

Theoretical saturation In the context of **qualitative** interviewing, the point of theoretical saturation is reached when additional **interviews** produce no evidence of attitudes and behaviour that have not already been encountered.

Topographic map A map showing patterns relating to a wide range of physical and human distributions, with an emphasis on tangible features.

Topology The relationship between spatially defined **entities**.

URL Uniform Resource Locator. The address which uniquely defines the location of any site on the **Internet**.

Validity For data to be valid they must not only be **reliable** but appropriate for the task in hand.

Variable A characteristic which varies in value from one case to another.

Variance An important **measure of dispersion** in which deviations from the mean are squared, summed, and divided by the number of occurrences. See **standard deviation**.

Vector data Data **storage** which allows screen displays to be constructed from stored sets of **co-ordinates** which define **entities**.

WAN Wide Area Network. A network built by establishing links between **LANs** (and also independent computers). Communication is typically routed by telephone and satellite links.

Web See **World Wide Web**.

Wilcoxon signed-rank test A **non-parametric test of difference** for two paired **samples**.

Window An information panel in a **GUI** such as Microsoft Windows.

Word-processor Software designed to let the user create and edit text, and present it at a reasonable standard of quality.

World Wide Web A rapidly growing and user-friendly area of the **Internet** distinguished by the use of images and **hypertext** rather than simple text and hierarchical searches.

WYSIWYG What You See Is What You Get. A term applied to software designed to produce a screen display identical in form to the final output.

Bibliography

Anderson, K. and Gale, F. (1992) *Inventing Places: Studies in Cultural Geography*, Melbourne: Longman Cheshire.

Anson, R.W. and Ormeling, F.J. (1988–1996) *Basic Cartography for Students and Technicians*, Oxford: Heinemann, 3 volumes and Exercise Manual.

Barber, P. and Board, C. (1993) *Tales from the Map Room*, London: BBC.

Bateson, N. (1984) *Data Construction in Social Surveys*, London: Allen & Unwin.

Bell, J. (1993) *Doing your Research Project*, Buckingham: Open University Press.

Blalock, H.M. (1979) *Social Statistics*, Tokyo: McGraw-Hill – Kogakusha, 2nd edition.

Buttenfield, B.P. and McMaster, R.B. (eds) (1991) *Map Generalization*, Harlow: Longman.

Cook, I. and Crang, M. (1995) *Doing Ethnographies* (CATMOG 58), Norwich: University of East Anglia.

Cosgrove, D., Roscoe, B., and Rycroft, S. (1996) 'Landscape and Identity at Ladybower Reservoir and Rutland Water', *Transactions of the Institute of British Geographers* New Series 21: 534–551.

Cresswell, T. (1993) 'Mobility as Resistance: A Geographical Reading of Kerouac's *On the Road*', *Transactions of the Institute of British Geographers* New Series 18: 249–262.

Crouch, D. and Matless, D. (1996) 'Refiguring Geography: Parish Maps of Common Ground', *Transactions of the Institute of British Geographers* New Series 21: 236–255.

Dale, A. and Marsh, C. (1993) *The 1991 Census User's Guide*, London: HMSO.

Daniels, S. and Rycroft, S. (1993) 'Mapping the Modern City: Alan Sillitoe's Nottingham Novels', *Transactions of the Institute of British Geographers* New Series 18: 460–480.

Davies, J. (1995) 'Chart Hits?', *Times Higher Educational Supplement*, 27 October: 19.

Denzin, N. and Lincoln, Y.S. (eds) (1994) *Handbook of Qualitative Research*, Thousand Oaks: Sage.

Department of the Environment (DoE) (1987) *Handling Geographic Information. Report of the Committee of Enquiry Chaired by Lord Chorley*, London: HMSO.

de Vaus, D.A. (1996) *Surveys in Social Research*, London: UCL Press, 4th edition.

Ebdon, D. (1985) *Statistics in Geography*, Oxford: Blackwell, 2nd edition.

Eyles, J. and Smith, D.M. (eds) (1988) *Qualitative Methods in Human Geography*, Cambridge: Polity Press.

Hahn, H. (1996) *The Internet: Complete Reference*, Berkeley: McGraw-Hill, 2nd edition.

Hakim, C. (1992) *Research Design, Strategies and Choices in the Design of Social Research*, London: Routledge.

Hammond, R. and McCullagh, P. (1978) *Quantitative Techniques in Geography: An Introduction*, Oxford: Clarendon, 2nd edition.

Haring, L.L., Lounsbury, J.F., and Frazier, J.W. (1992) *Introduction to Scientific Geographic Research*, Dubuque: W.C. Brown, 4th edition.

Harley, J.B. (1975) *Ordnance Survey Maps: A Descriptive Manual*, Southampton: Ordnance Survey.

Harré, R. (1979) *Social Being*, Oxford: Blackwell.

Harvey, D. (1969) *Explanation in Geography*, London: Arnold.

Hay, I. (1996) *Communication in Geography and the Environmental Sciences*, Oxford: Oxford Unversity Press.

Hickin, W., Maguire, D.J., and Strachan, A.J. (1991) *Introduction to GIS: The Arc/Info Method*, Leicester: Midlands Regional Research Laboratory.

Jackson, P. (1989) *Maps of Meaning*, London: Routledge.

Keates, J. (1989) *Cartographic Design and Production*, Harlow: Longman, 2nd edition.

Keates, J. (1996) *Understanding Maps*, Harlow: Longman, 2nd edition.

Kraak, M. J. and Ormeling, F. J. (1996) *Cartography – Visualization of Spatial Data*, Harlow: Longman.

Krol, E. (1994) *The Whole Internet User's Guide and Catalog*, Sebastopol: O'Reilly, 2nd edition.

Leslie, S. (1996) 'Convergent Thinking Needed', *Mapping Awareness* 10 (5): 13.

MacEachran, A.M. (1995) *How Maps Work: Representation, Visualization, and Design*, New York: Guilford Press.

MacEachran, A.M. and Taylor, D.R.F. (eds) (1994) *Visualization in Modern Cartography*, Oxford: Pergamon.

McGinn, J. (1996) 'Geography Lessons', *Computing* 11 July: 30–31.

Maguire, D. (1989) *Computers in Geography*, Harlow: Longman.

Maguire, D. (1991) 'An Overview and Definition of GIS', in Maguire, D.,

Goodchild, M.F., and Rhind, D.W. (eds) *Geographical Information Systems*, Harlow: Longman, 2 volumes.

Maguire, D. and Dangermond, J. (1991) 'The Functionality of GIS', in Maguire, D., Goodchild, M.F., and Rhind, D.W. (eds) *Geographical Information Systems*, Harlow: Longman, 2 volumes.

Maguire, D., Goodchild, M.F., and Rhind, D.W. (eds) (1991) *Geographical Information Systems*, Harlow: Longman, 2 volumes.

Marshall, C. and Rossman, G.B. (1995) *Designing Qualitative Research*, Thousand Oaks: Sage, 2nd edition.

Massey, D. and Jess, P. (eds) (1995) *A Place in the World?*, Milton Keynes: Open University/Oxford University Press.

Mather, P. (1991) *Computer Applications in Geography*, Chichester: Wiley.

May, J. (1996) 'Globalization and the Politics of Place: Place and Identity in an Inner London Neighbourhood', *Transactions of the Institute of British Geographers* New Series 213: 194–215.

Miles, M.B. and Huberman, A.M. (1994) *Qualitative Data Analysis*, Thousand Oaks: Sage, 2nd edition.

Monk, J. (1992) 'The Construction and Deconstruction of Women's Roles', pp. 123–156 in K. Anderson and F. Gale (eds) *Inventing Places: Studies in Cultural Geography*, Melbourne: Longman Cheshire.

Moser, C. and Kalton, G. (1971) *Survey Methods in Social Investigation*, London: Heinemann, 2nd edition.

Northedge, A. (1990) *The Good Study Guide*, Milton Keynes: Open University Press.

O'Brien, L. (1992) *Introduction to Quantitative Geography*, London: Routledge.

Oppenheim, A.N. (1992) *Questionnaire Design, Interviewing, and Attitude Measurement*, London: Pinter.

Parsons, A. and Knight, P. (1995) *How to Do your Dissertation in Geography and Related Disciplines*, London: Chapman & Hall.

Pope, I. (1995) *Internet UK*, Hemel Hempstead: Prentice Hall International.

Robinson, A., Morrison, J., Muercke, P.C., Guptill, S.C., and Kimerling, A.J. (1994) *Elements of Cartography*, New York: Wiley, 6th edition.

Rose, G. (1994) 'The Cultural Politics of Place: Local Representation and Oppositional Discourse in Two Films', *Transactions of the Institute of British Geographers* New Series 19: 46–60.

Sayer, A. (1992) *Method in Social Science*, Routledge: London, 2nd edition.

Schofield, S. (1995) *The UK Internet Book*, Wokingham: Addison Wesley.

Siegel, S. and Castellan, N.J., Jr (1988) *Nonparametric Statistics for the Behavioral Sciences*, New York: McGraw-Hill.

Silverman, D. (1993) *Interpreting Qualitative Data*, London: Sage.

Smith, S.J. (1993) 'Bounding the Borders; Claiming Space and Making Place in Rural Scotland', *Transactions of the Institute of British Geographers* New Series 18: 291–308.

Toyne, P. and Newby, P.T. (1971) *Techniques in Human Geography*, London: Macmillan.

Wood, D. (1993) *The Power of Maps*, London: Routledge.

Index

Printed and bound by CPI Group (UK) Ltd, Croydon, CR0 4YY

01/11/2024

01782610-0016